Planning with Markov Decision Processes

An AI Perspective

Synthesis Lectures on Artificial Intelligence and Machine Learning

Editors
Ronald J. Brachman, *Yahoo! Research*
William W. Cohen, *Carnegie Mellon University*
Thomas Dietterich, *Oregon State University*

Planning with Markov Decision Processes: An AI Perspective

Mausam and Andrey Kolobov

ISBN: 978-3-031-00431-5   paperback
ISBN: 978-3-031-01559-5   ebook

DOI 10.1007/978-3-031-01559-5

A Publication in the Springer series
SYNTHESIS LECTURES ON ARTIFICIAL INTELLIGENCE AND MACHINE LEARNING

Lecture #17
Series Editors: Ronald J. Brachman, *Yahoo Research*
                William W. Cohen, *Carnegie Mellon University*
                Thomas Dietterich, *Oregon State University*
Series ISSN
Synthesis Lectures on Artificial Intelligence and Machine Learning
Print 1939-4608   Electronic 1939-4616

Planning with
Markov Decision Processes

An AI Perspective

Mausam and Andrey Kolobov
University of Washington

SYNTHESIS LECTURES ON ARTIFICIAL INTELLIGENCE AND MACHINE LEARNING #17

ABSTRACT

Markov Decision Processes (MDPs) are widely popular in Artificial Intelligence for modeling sequential decision-making scenarios with probabilistic dynamics. They are the framework of choice when designing an intelligent agent that needs to act for long periods of time in an environment where its actions could have uncertain outcomes. MDPs are actively researched in two related subareas of AI, probabilistic planning and reinforcement learning. Probabilistic planning assumes known models for the agent's goals and domain dynamics, and focuses on determining how the agent should behave to achieve its objectives. On the other hand, reinforcement learning additionally learns these models based on the feedback the agent gets from the environment.

This book provides a concise introduction to the use of MDPs for solving probabilistic planning problems, with an emphasis on the algorithmic perspective. It covers the whole spectrum of the field, from the basics to state-of-the-art optimal and approximation algorithms. We first describe the theoretical foundations of MDPs and the fundamental solution techniques for them. We then discuss modern optimal algorithms based on heuristic search and the use of structured representations. A major focus of the book is on the numerous approximation schemes for MDPs that have been developed in the AI literature. These include determinization-based approaches, sampling techniques, heuristic functions, dimensionality reduction, and hierarchical representations. Finally, we briefly introduce several extensions of the standard MDP classes that model and solve even more complex planning problems.

KEYWORDS

MDP, AI planning, probabilistic planning, uncertainty in AI, sequential decision making under uncertainty, reinforcement learning

To my grandmother, Mrs. Shanti Agrawal,
who inspired a culture of authorship in the family.

— Mausam

To Nina Kolobova and Vadim Kolobov,
the best parents anyone could ask for.

— Andrey

To Prof. Daniel S. Weld,
who guided us into the field of MDP planning.

— Mausam & Andrey

Contents

Preface

Starting in the 1950s, over time, a number of books have been written on Markov Decision Processes in different fields of research. However, most books have taken a rather theoretical view of the topic – delving deep into the fundamental insights, elaborating on the underlying mathematical principles and proving each detail to give the framework the theoretical consistency that is expected of such a model.

Comparatively, there is less synthesized literature available on the use of MDPs within AI. While the reinforcement learning perspective has been published in a couple of books and surveys, the lack of surveys is especially glaring from the probabilistic planning point of view.

Our book differs from the existing literature on MDPs in other fields with its emphasis on the *algorithmic* techniques. We start with the fundamental algorithms, but go far beyond them and survey the multitude of approaches proposed by AI researchers in scaling the solution algorithms to larger problems. Wherever necessary, we do present the theoretical results, but in line with our focus, we avoid the proofs and point to other literature for an in-depth analysis.

We make no assumptions about the reader's prior knowledge of MDPs, and expect this book to be of value to a beginning student interested in learning about MDPs, to a mid-level researcher who wishes to get an overview of the MDP solution techniques, as well as to a seasoned researcher interested in the references to advanced techniques from where she can launch further study.

Our book comprises seven chapters. We provide a general introduction to the book and MDPs in the first chapter. Chapter 2 defines the representation of an MDP model. There is an important issue here. Various researchers have studied slightly different problem definitions of an MDP and not all results and algorithms apply to all versions. To keep the book coherent, we chose the stochastic shortest path (SSP) formalism as our base MDP. This generalizes several common MDP models, but, some other advanced models are out of its purview.

After the model definition, we focus the next three chapters on the *optimal* solution algorithms. Chapter 3 starts from the fundamental algorithms from the 1950s and leads up to a variety of recent optimizations that scale them. Heuristic search ideas from the AI literature are incorporated on top of these algorithms in Chapter 4. These ideas are useful when a specific start state is known to us. Chapter 5 describes the use of compact value function representations in optimal MDP algorithms.

An optimal solution to an MDP is a luxury. Most real problems are so large that computing optimal solutions is infeasible. A significant emphasis of our book is Chapter 6, which discusses the state of the art in approximately solving these MDPs. Researchers have proposed a wide range of algorithms, which span various points on the efficiency-optimality spectrum. This chapter surveys these algorithms.

Finally, Chapter 7 briefly discusses several models that relax the various assumptions implicit in an MDP. Continuous and hybrid state spaces, concurrent, durative actions, and generalizations of SSPs are some of the MDP models that we discuss in the last chapter.

We would like to acknowledge several colleagues and researchers who gave us useful feedback on earlier drafts of the book or helped us with specific concepts. These include Chris Lin (University of Washington), Dan Weld (University of Washington), Dimitri Bertsekas (Massachusetts Institute of Technology), Eric Hansen (Mississippi State University), Hector Geffner (Universitat Pompeu Fabra), Martine De Cock (Ghent University), Peng Dai (Google), Scott Sanner (National ICT Australia Ltd), Florent Teichteil-Königsbuch (ONERA), and the anonymous reviewer. We also thank Tom Dietterich (Oregon State University) for his enthusiasm on the initial idea and Mike Morgan for his constant help while writing the book.

The writing of this book has been supported by the NSF grant IIS-1016465, ONR grant N000140910051, and Turing Center at the University of Washington. Any opinions or conclusions expressed in the text are those of the authors and do not necessarily reflect the views of the funding agencies.

Mausam and Andrey Kolobov
June 2012

CHAPTER 1

Introduction

The vision of artificial intelligence is often manifested through an autonomous agent in a complex and uncertain environment. The agent is capable of thinking ahead and acting for long periods of time in accordance with its goal/objective. Such agents appear in a broad set of applications, for example, the Mars rover planning its daily schedule of activities [166], planning of military operations [2], robocup soccer [223], an agent playing games like blackjack [195], a set of elevators operating in sync [57], and intervention of cellular processes [47].

The AI sub-field of *Automated Planning under Uncertainty* tackles several core problems in the design of such an agent. These planning problems are typically formulated as an instance of a *Markov Decision Process* or an MDP. At the highest level, an MDP comprises a set of world states, a set of actions under the agent's control, a transition model describing the probability of transitioning to a new state when taking an action in the current state, and an objective function, e.g., maximizing the sum of rewards obtained over a sequence of time steps. An MDP solution determines the agent's actions at each decision point. An optimal MDP solution is one that optimizes the objective function.

The MDP model was popularized in the operations research literature with the early works of Bellman and Howard in 1950s [18; 114]. For the first thirty or so years after this, there was significant progress in building the theory of MDPs and a basic set of algorithms to solve them [83]. The AI community adopted the model in the early 1990s, with the earliest works exploring connections to the popular, classical planning paradigm (which ignores the uncertainty in action outcomes) [36; 70; 74; 129].

Since then, MDPs have been immensely popular in AI in primarily two related sub-communities – probabilistic planning and reinforcement learning. The probabilistic planning literature assumes complete prior knowledge of the MDP model and focuses on developing computationally efficient approaches to solve it. On the other hand, reinforcement learning studies the harder problem in which the agent does not have prior access to the complete model and, hence also has to learn (parts of) it based on its experience. In this book, we survey the state-of-the-art techniques developed in the probabilistic planning literature.

1.1 CHARACTERISTICS OF AN MDP

What kinds of domains are best modeled as an MDP? We answer the question below and use two scenarios as our running examples.

The first example is of an agent playing the game of blackjack. The goal of the game is to acquire playing cards such that the sum of the scores is higher than the dealer's cards, but not over

21. Here a state comprises the known cards of the agent as well as the dealer, the actions are agent's choices like asking for a new card (in blackjack's terms: hit, stand, split, etc.).

Another example scenario is the planning of daily activities for NASA's rover on Mars. The rover is autonomous for the most part of the day due to a limited communication bandwidth. The goals for the rover are set by the scientists on Earth. Its state space includes its current location, battery life, current time of the day, etc. Its actions comprise moving to another location, performing various experiments on different rocks, and so on. Each experiment may lead to scientific payoffs, so those are a part of the rover's objective function, as well as the fact that it should stay safe by not attempting any dangerous actions.

We now discuss the key characteristics of these domains that are conveniently modeled in an MDP.

Uncertain Domain Dynamics. Although people have studied deterministic MDPs also, the framework is most valuable when the domain has uncertainty. The uncertainty may be in the agent's own actions or in the exogenous factors in the domain. For example, in blackjack, asking for a new card has a stochastic effect – it can yield any card with a known probability distribution. Similarly, the rover on Mars may try to move forward one foot, but, due to uncertainty in the motors' speed, it may end up moving by a slightly different amount.

Sequential Decision Making. MDPs are the model of choice when the agent needs to take a *sequence* of actions and thus needs to plan ahead. The benefit of an action may not be immediately clear to a myopic agent; however, an MDP-based agent may prefer that action if it helps achieve a higher *future* payoff (or a lower future cost). As an example, an MDP-based rover may choose to move a long way to another location to achieve a higher reward, instead of spending time on the low-reward rocks that are nearer. Even though the travel has a high upfront cost, it is better in the long run. In that sense, MDPs are a principled mechanism to trade off multiple competing objectives (like cost-reward trade-offs).

Cyclic Domain Structures. MDPs and associated algorithms are especially valuable when a state may be revisited in a domain, i.e., the state space is cyclic in the graph-theoretic sense (we also discuss acyclic MDPs in this book). These commonly occur in real domains. Imagine the rover trying to pick up a small rock. It may fail in picking it up or the rock may fall down taking the rover back to the same world state. Such spaces are handled easily by the MDP model.

Nature is Fair. MDPs best model the scenarios where the outcomes happen by *chance* and are not decided by an adversary. In other words, the sampling of the next state is performed by a fair nature, based on a known probability distribution that depends only on the agent's state and action. For example, the blackjack dealer is not purposely trying to defeat the player and the next card is just sampled at random from a fair deck.

Full Observability and Perfect Sensors. MDPs assume that the whole of the state space is visible to the agent and its sensors are perfect. For example, the blackjack agent can see all the open cards and its sensors do not make a mistake in reading the cards. Problems with noisy sensors are modeled as Partially Observable MDPs (POMDPs). Those are not our focus, and we briefly discuss them in Section 7.5.

1.2   CONNECTIONS WITH DIFFERENT FIELDS

The MDP framework has close connections to a wide variety of fields. We list some of these below.

Operations Research.   OR was responsible for the genesis of the model. Some of the earliest applications were in inventory control and targeted advertising. It is said that SEARS profited millions of dollars in the 1960s, when an MDP-based decision-maker chose the subset of customers to whom the catalog was sent by post [116]. Since then, MDPs have seen innumerable applications coming out of the OR community [240].

Computational Finance.   Finance is another perfect subarea for the application of MDPs, since the market fluctuates, and the decision-making process needs to incorporate long-term rewards. As an example, a credit-card company managed its customers' credit-lines using an MDP and made profits worth $75 million per year [234]. For a survey of finance applications using MDPs please read this book [14].

Control Theory.   Electrical/mechanical engineers designing automated physical systems are faced with the algorithmic task of controlling their systems. Here, the system dynamics capture the laws of physics, e.g., characteristics of the motors running the systems [148; 222]. Such problems are commonly studied in control theory and use the MDP formalism as their basis.

Computational Neuroscience.   MDPs and extensions are popular in computational neuroscience for formalizing or understanding the process of decision-making under uncertainty in humans and animals [202]. Behavioral psychology has strong connections with reinforcement learning, which is a field of AI based on MDPs.

Gambling.   Gambling theory [81] is also closely connected to MDPs. Like our blackjack example, several gambling problems are games against chance trying to maximize a long-term utility, a perfect fit for MDPs.

Graph-Theory.   An MDP domain description defines an AND-OR graph. The AND-OR graphs are a generalization of directed graphs. They allow two kinds of nodes – the OR nodes in which the agent chooses an edge, and the AND node in which nature chooses an edge. An MDP algorithm solves the stochastic shortest path problem in cyclic AND-OR graphs.

Statistics.   While not very popular in the Statistics literature, an MDP is closely related to a Markov chain. A Markov chain specifies a state-transition model for a system that is transitioning of its own accord, *without* an agent's intervention. An MDP, on the other hand, models the scenario when an agent's actions affect the system's transitions. In this view, an MDP can be thought of as a collection of Markov chains and has a long-term reward associated with each of them. The goal of the agent is to choose the best Markov chain for execution, one that maximizes the long-term reward.

Artificial Intelligence.   AI, which is also the focus of this book, has used MDPs extensively in designing rational agents acting for long periods of time. Robotics researchers use MDPs in modeling intelligent robots. The field of Game Playing studies an agent playing a game. Whenever the game's domain has uncertainty (e.g., rolling a die) and is against nature (as opposed to adversary) MDPs are the tool of choice. Finally, machine learning studies the reinforcement learning formalism that focuses on learning the MDP model while also acting in the world.

1.3   OVERVIEW OF THIS BOOK

Starting in the 1950s, over time, a number of books have been written on Markov Decision Processes in different fields of research [21; 24; 197]. However, most books have taken a rather theoretical view of the topic – delving deep into the fundamental insights, elaborating on the underlying mathematical principles and proving each detail to give the framework the theoretical consistency that is expected of such a model.

Comparatively, there is less synthesized literature available on the use of MDPs within AI. While the reinforcement learning perspective has been published in a couple of books and surveys [121; 224; 226], the lack of surveys is especially glaring from the probabilistic planning point of view.

Our book contrasts with the existing literature on MDPs in other fields with its emphasis on the *algorithmic* techniques. We start with the fundamental algorithms, but go much beyond them and survey the multitude of approaches proposed by AI researchers in scaling the solution algorithms to larger problems. Wherever necessary, we do present the theoretical results, but in line with our focus, we avoid the proofs and point to other literature for an in-depth analysis.

We make no assumptions on the reader's prior knowledge of MDPs, and expect this book to be of value to a beginning student interested in learning about MDPs, to a mid-level researcher who wishes to get an overview of the MDP solution techniques, as well as to a seasoned researcher interested in the references to advanced techniques from where she can launch a further study.

Our book comprises six further chapters. Chapter 2 defines the representation of an MDP model. There is an important issue here. Various researchers have studied slightly different problem definitions of an MDP and not all results and algorithms apply to all versions. To keep the book coherent, we choose the stochastic shortest path (SSP) formalism as our base MDP. This generalizes several common MDP models, but, some other advanced models are out of its purview. We briefly discuss extensions to SSPs in Section 7.4.

After the model definition, we focus the next three chapters on the *optimal* solution algorithms. Chapter 3 starts from the fundamental algorithms from the 1950s and leads up to a variety of recent optimizations that scale them. Heuristic search ideas from the AI literature are incorporated on top of these algorithms in Chapter 4. These ideas are useful when a specific start state is known to us. Chapter 5 describes the use of compact value function representations in optimal MDP algorithms.

An optimal solution to an MDP is a luxury. Most real problems are so large that computing optimal solutions is infeasible. A significant emphasis of our book is Chapter 6, which discusses the state of the art in approximately solving these MDPs. Researchers have proposed a wide range of algorithms, which span various points on the efficiency-optimality spectrum. This chapter surveys these algorithms.

Finally, Chapter 7 briefly discusses several models that relax the various assumptions implicit in an MDP. Continuous and hybrid state spaces, concurrent, durative actions, and generalizations of SSPs are some of the MDP models that we discuss in the last chapter.

CHAPTER 2

MDPs

Under its broadest definition, the concept of an MDP covers an immense variety of models. However, such generality comes at a cost. The definition provides too little structure to make it useful for deriving efficient MDP solution techniques. Instead, practitioners in AI and other areas have come up with more specialized but also more structured MDP classes by adding restrictions to the basic MDP notion.

We find the process of identifying interesting MDP types by narrowing down an abstract MDP description to be very instructive, and lay out this chapter as follows. First, using examples from Chapter 1, we formulate a very general MDP definition (although still not the most general one possible!). Next, we analyze its weaknesses and show ways to fix them by adding more conditions to the definition while making sure the definition continues to cover an interesting class of real-world scenarios. In this way, we derive stochastic shortest path (SSP) MDPs, the class we will concentrate on in the remainder of the book. We hope that the reader will complete the chapter with a good understanding of both *why* SSPs are defined as they are, and *how* different parts of their definition determine their mathematical properties.

2.1 MARKOV DECISION PROCESSES: DEFINITION

The example applications from the previous chapter have several common characteristics:

- Each of the described scenarios involves a system that moves through a sequence of *states*. E.g., in blackjack the state of the system is described by which cards are in the deck, which have been dealt to the dealer, and which have been dealt to the player.

- The transitions between states happen in response to a sequence of *actions* taken by an agent at various *decision epochs*. In blackjack, decision epochs are rounds in the game. In each round, the player's (agent's) choice of actions is to take another card ("hit"), not to take any more cards ("stand"), etc.

- At any decision epoch, the state of the system is fully known to the agent. In many versions of blackjack, the player knows the dealer's cards, his own cards, and can infer the current composition of the deck. (Although the latter is considered "card counting" and is typically disallowed in casinos, the player does, in principle, have access to this information.)

- The outcomes of actions are typically not deterministic. An action executed in a state can lead the system to several states with different *transition probabilities* known to the agent. Again using the blackjack example, if the player opts to get another card from the deck, he can get any card with equal probability. Assuming the player counted cards, he knows what cards are in the deck and hence the probability of getting any particular one of them.

- Executing an action on the system brings some *reward*. In case executing an action requires some payment of some resource, the reward is negative and is more appropriately viewed as *cost*. Moreover, the agent knows the reward/cost for each possible outcome before executing the action. Blackjack has a peculiar reward structure. No action except for the last one gives the player any reward or incurs any cost. Such scenarios are said to involve *delayed rewards*. In the Mars rover example, most actions do have an immediate reward or cost — the cost of materials for conducting an experiment, the reward for a successful experiment's completion, and so on.

- The *objective* is to control the system by taking appropriate actions while optimizing for some criterion. The optimization criterion can be, for instance, maximizing the total expected reward over some sequence of decision epochs, although there are plenty of other possibilities. In the case of blackjack, the total expected reward is the expected win amount.

Not all MDPs studied in the literature have these properties. The field of reinforcement learning studies processes in which the actions' outcome probabilities are not available *a priori*. Partially Observable MDPs (POMDPs) do away with the assumption of perfect knowledge of the current state. Different communities have researched other kinds of models as well; we discuss them briefly in Section 7.5. However, in most of the book we primarily examine MDPs that conform to the assumptions above and a few more, as stated in the following definition:

Definition 2.1   Finite Discrete-Time Fully Observable Markov Decision Process (MDP). A finite discrete-time fully observable MDP is a tuple $\langle \mathcal{S}, \mathcal{A}, \mathcal{D}, \mathcal{T}, \mathcal{R} \rangle$, where:

- $\mathcal{S}$ is the finite set of all possible states of the system, also called the *state space*;

- $\mathcal{A}$ is the finite set of all actions an agent can take;

- $\mathcal{D}$ is a finite or infinite sequence of the natural numbers of the form $(1, 2, 3, \ldots, T_{max})$ or $(1, 2, 3, \ldots)$ respectively, denoting the decision epochs, also called *time steps*, at which actions need to be taken;

- $\mathcal{T} : \mathcal{S} \times \mathcal{A} \times \mathcal{S} \times \mathcal{D} \to [0, 1]$ is a transition function, a mapping specifying the probability $\mathcal{T}(s_1, a, s_2, t)$ of going to state s_2 if action a is executed when the agent is in state s_1 at time step t;

- $\mathcal{R} : \mathcal{S} \times \mathcal{A} \times \mathcal{S} \times \mathcal{D} \to \mathbb{R}$ is a reward function that gives a finite numeric reward value $\mathcal{R}(s_1, a, s_2, t)$ obtained when the system goes from state s_1 to state s_2 as a result of executing action a at time step t.

Since we will almost exclusively discuss finite discrete-time fully observable MDPs as opposed to any others (the only exception being Chapter 7), we will refer to them simply as "MDPs" in the rest of the book.

Notice that mappings $\mathcal{T}$ and $\mathcal{R}$ may depend on the number of time steps that have passed since the beginning of the process. However, neither $\mathcal{T}$ nor $\mathcal{R}$ depends on the sequence of states the system has gone through so far, instead depending only on the state the system is in currently. This independence assumption is called the *first-order Markov assumption* and is reflected in the MDPs' name.

2.2   SOLUTIONS OF AN MDP

Intuitively, solving an MDP means finding a way of choosing actions to control it. Recall that MDPs' actions typically have several possible outcomes. An agent cannot pick a state to which the system will transition when the agent executes a particular action, since the exact outcome is not in the agent's control. Therefore, after executing a sequence of actions we may find the system in one of many possible states. For instance, after several rounds in blackjack, the player's and the dealer's hands can be virtually any combination of cards. To be robust in such situations, our method of picking actions should enable us to decide on an action no matter which state we are in. Thus, what we need is a global *policy*, a rule for action selection that works in any state.

What information might such a rule use? The knowledge of the current state can clearly be very useful. In general, however, we may want our decision to depend not only on the current state but on the entire sequence of states the system has been through so far, as well as the sequence of actions we have chosen up to the present time step, i.e., the entire *execution history*.

Definition 2.2   Execution History. An *execution history* of an MDP up to time step $t \in \mathcal{D}$ is a sequence $h_t = ((s_1, a_1), \ldots, (s_{t-1}, a_{t-1}), s_t)$ of pairs of states the agent has visited and actions the agent has chosen in those states for all time steps t' s.t. $1 \le t' \le t - 1$, plus the state visited at time step t.

We denote the set of all possible execution histories up to decision epoch t as $\mathcal{H}_t$, and let $\mathcal{H} = \cup_{t \in \mathcal{D}} \mathcal{H}_t$.

Note also that our rule for selecting actions need not be deterministic. For instance, when faced with a choice of several equally good actions, we might want to pick one of them at random in order to avoid a bias. Thus, in the most general form, a solution policy for an MDP may be not only history-dependent but also probabilistic.

Definition 2.3   History-Dependent Policy. A *probabilistic history-dependent policy* for an MDP is a probability distribution $\pi : \mathcal{H} \times \mathcal{A} \to [0, 1]$ that assigns to action $a \in \mathcal{A}$ a probability $\pi(h, a)$ of choosing it for execution at the current time step if the execution history up to the current time step is $h \in \mathcal{H}$. A *deterministic history-dependent policy* is a mapping $\pi : \mathcal{H} \to \mathcal{A}$ that assigns to each

$h \in \mathcal{H}$ an action $a \in \mathcal{A}$ to be executed at the current time step if h the execution history up to the current time step.

It is easy to see that a deterministic history-dependent policy is a probabilistic policy that, for every history, assigns the entire probability mass to a single action. Accordingly, the deterministic policy notation $\pi(h) = a$ is just a shorthand for the probabilistic policy notation $\pi(h, a) = 1$.

One can view the process of executing a history-dependent policy π as follows. At every time step, the agent needs to examine the history h of past states and action choices up to that time step, and determine the distribution over actions that corresponds to h under π. To get an actual action recommendation from π, the agent needs to sample an action from that distribution (if π is deterministic, this step is trivial). Executing the sampled action brings the agent to another state, and the procedure repeats.

While very general, this definition suggests that many MDP solutions may be very hard to compute and represent. Indeed, a history-dependent policy must provide a distribution over actions for every possible history. If the number of time steps $|\mathcal{D}|$ in an MDP is infinite, its number of possible histories is infinite as well. Thus, barring special cases, solving an MDP seemingly amounts to computing a function over an infinite number of inputs. Even when $\mathcal{D}$ is a finite set, the number of histories, and hence the maximum size of an MDP solution, although finite, may grow exponentially in $|\mathcal{D}|$.

Due to the difficulties of dealing with arbitrary history-dependent policies, the algorithms presented in this book aim to find more compact MDP solutions in the form of *Markovian policies*.

Definition 2.4   **Markovian Policy.** A probabilistic (deterministic) history-dependent policy $\pi :$ $\mathcal{H} \times \mathcal{A} \rightarrow [0, 1] \, (\pi : \mathcal{H} \rightarrow \mathcal{A})$ is *Markovian* if for any two histories $h_{s,t}$ and $h'_{s,t}$, both of which end in the same state s at the same time step t, and for any action a, $\pi(h_{s,t}, a) = \pi(h'_{s,t}, a) \, (\pi(h_{s,t}) = a$ if and only if $\pi(h'_{s,t}) = a)$.

In other words, the choice of an action under a Markovian policy depends only on the current state and time step. To stress this fact, we will denote probabilistic Markovian policies as functions $\pi : \mathcal{S} \times \mathcal{D} \times \mathcal{A} \rightarrow [0, 1]$ and deterministic ones as $\pi : \mathcal{S} \times \mathcal{D} \rightarrow \mathcal{A}$.

Fortunately, disregarding non-Markovian policies is rarely a serious limitation in practice. Typically, we are not just interested in finding *a* policy for an MDP. Rather, we would like to find a "good" policy, one that optimizes (or nearly optimizes) some objective function. As we will see shortly, for several important types of MDPs and objective functions, at least one optimal solution *is* necessarily Markovian. Thus, by restricting attention only to Markovian policies we are not foregoing the opportunity to solve these MDPs optimally.

At the same time, dropping non-Markovian history-dependent policies from consideration makes the task of solving an MDP much easier, as it entails deciding on a way to behave "merely" for every state and time step. In particular, if $\mathcal{D}$ is finite, the size of a policy specification is at most linear in $|\mathcal{D}|$. Otherwise, however, the policy description size may still be infinite. To address this, for MDPs with an infinite number of steps we narrow down the class of solutions even further by concentrating only on *stationary* Markovian policies.

Definition 2.5   Stationary Markovian Policy. A probabilistic (deterministic) Markovian policy $\pi : S \times D \times A \to [0, 1]$ $(\pi : S \times D \to A)$ is *stationary* if for any state s, action a, and two time steps t_1 and t_2, $\pi(s, t_1, a) = \pi(s, t_2, a)$ $(\pi(s, t_1) = a$ if and only if $\pi(s, t_2) = a)$, i.e., π does not depend on time.

Since the time step plays no role in dictating actions in stationary Markovian policies, we will denote probabilistic stationary Markovian policies as functions $\pi : S \times A \to [0, 1]$ and determin- istic ones as $\pi : S \to A$. Stationary solutions look feasible to find — they require constructing an action distribution for every state and hence have finite size for any MDP with a finite state space. However, they again raise the concern of whether we are missing any important MDP solutions by tying ourselves only to the stationary ones. As with general Markovian policies, for most practically interesting MDPs with an infinite number of time steps this is not an issue, because they have at least one best solution that is stationary.

In the discussion so far, we have loosely used the terms "best" and "optimal" to indicate the quality of an MDP policy. Now that we have defined what an MDP solution is, it is only natural to ask: precisely how can we compare policies to each other, and on what basis should we prefer one policy over another? Clearly, it helps to associate a numeric value to each policy, but what would this value measure be? When executing a policy, i.e., applying actions recommended by it in various states, we can expect to get associated rewards. Therefore, it makes intuitive sense to prefer a policy that controls the MDP in a way that maximizes some *utility function* of the collected reward sequence, i.e., define the value of a policy to be the policy's utility. For now, we intentionally leave this utility function unspecified and first formalize the concept of a policy's value.

Definition 2.6   Value Function. A *history-dependent value function* is a mapping $V : \mathcal{H} \to [-\infty, \infty]$. A *Markovian value function* is a mapping $V : S \times D \to [-\infty, \infty]$. A *stationary Marko- vian value function* is a mapping $V : S \to [-\infty, \infty]$.

The Markovian value function notation $V : S \times D \to [-\infty, \infty]$ is just syntactic sugar for a history-dependent value function $V : \mathcal{H} \to [-\infty, \infty]$, that, for all pairs of histories $h_{s,t}$ and $h'_{s,t}$ that terminate at the same state at the same time, has $V(h_{s,t}) = V(h'_{s,t})$. In other words, for such a history-dependent value function, $V(s, t) = V(h_{s,t})$ for all policies $h_{s,t}$. Analogously, if a Markovian policy has $V(s, t) = V(s, t')$ for states s and for all pairs of time steps t, t', then $V(s)$ is the value denoting $V(s, t)$ for any time step t. We will use $V(s, t)$ and $V(s)$ as the shorthand notation for a value function wherever appropriate.

Definition 2.7   The Value Function of a Policy. Let $h_{s,t}$ be a history that terminates at state s and time t. Let $R_{t'}^{\pi_{h_{s,t}}}$ be random variables for the amount of reward obtained in an MDP as a result of executing policy π starting in state s for all time steps t' s.t. $t \leq t' \leq |\mathcal{D}|$ if the MDP ended up in state s at time t via history $h_{s,t}$. *The value function* $V^\pi : \mathcal{H} \rightarrow [-\infty, \infty]$ *of a history-dependent policy* π *is a utility function* u *of the reward sequence* $R_t^{\pi_{h_{s,t}}}, R_{t+1}^{\pi_{h_{s,t}}}, \ldots$ *that one can accumulate by executing* π *at time steps* $t, t+1, \ldots$ *after history* $h_{s,t}$. Mathematically, $V^\pi(h_{s,t}) = u(R_t^{\pi_{h_{s,t}}}, R_{t+1}^{\pi_{h_{s,t}}}, \ldots)$.

While this definition is somewhat involved, it has a simple meaning. It says that the value of a policy π is the amount of utility we can expect from executing π starting in a given situation, whatever we choose our utility to be. We stress once more that the exact way of converting a sequence of rewards into utility and hence into the value of a policy is still to be made concrete.

The notation for the policy value function is simplified for Markovian and stationary Markovian policies π. In the former case, $V^\pi(s, t) = u(R_t^{\pi_{s,t}}, R_{t+1}^{\pi_{s,t}}, \ldots)$, and in the latter case $V^\pi(s) = u(R_t^{\pi_s}, R_{t+1}^{\pi_s}, \ldots)$. As already mentioned, for most MDP classes we will study in this book, the optimal solution will be a stationary Markovian policy, so our algorithms will concentrate on this type of policy when solving MDPs. Thus, most often we will be using the lightest notation $V^\pi(s) = u(R_t^{\pi_s}, R_{t+1}^{\pi_s}, \ldots)$.

The above definition of a policy's value finally allows us to formulate the concept of an *optimal MDP solution*.

Definition 2.8 Optimal MDP Solution. An *optimal solution* to an MDP is a policy π^* s.t. the value function of π^*, denoted as V^* and called *the optimal value function*, dominates the value functions of all other policies for all histories h. Mathematically, for all $h \in \mathcal{H}$ and any π, V^* must satisfy $V^*(h) \geq V^\pi(h)$.

In other words, given an optimality criterion (i.e., a measure of how good a policy is, as determined by the utility function u in Definition 2.7), an optimal MDP solution is a policy π^* that is at least as good as any other policy in every state according to that criterion. *An optimal policy is one that maximizes a utility of rewards.*

So far, we have examined MDPs in a very general form. This form allows us to precisely formulate an extremely wide range of scenarios and solutions to them. In the subsequent sections of this chapter, we will see that this flexibility comes at a steep premium. A practically interesting MDP needs to have an optimal solution that is not only defined but actually *exists*. Moreover, it should lend itself to reasonably efficient solution techniques. As we demonstrate in the next section, MDPs as presented in Section 2.1 generally fail to have either of these properties. We then formulate several condition sets s.t. for any MDP satisfying any of these condition sets, an optimal policy is guaranteed to exist. By imposing additional restrictions on the MDP definition (2.1), we gradually derive several classes of MDPs for which this policy can be found in practice.

2.3 SOLUTION EXISTENCE

Definition 2.8 contains two caveats. First, it leaves the utility function u, which determines how a sequence of random variables denoting rewards is converted into a utility, completely unspecified.

In the meantime, one needs to be very careful when defining it. Consider the following seemingly natural choice for deriving utility from a sequence of arbitrary random variables denoting rewards: $u(R_t, R_{t+1}, \ldots) = \sum_{t'=t}^{|\mathcal{D}|} R_{t'}$. Under this u, the value of a policy π would be the sum of rewards the execution of π yields. Unfortunately, such a choice of u is not a well-defined function. MDPs' actions generally cause probabilistic state transitions, as described by the transition function $\mathcal{T}$, so the sequence of states the agent goes through by following π from a given state (or, after a given history) differs across π's executions. As a result, the sequence of rewards collected each time may be different as well.

The second caveat concerns the possibility of comparing value functions for two different policies. Each V^π is, in effect, a vector in a Euclidean space — the dimensions correspond to histories/states with values V^π assigns to them. Thus, comparing quality of policies is similar to comparing vectors in $\mathbb{R}^m$ componentwise. Note, however, that vectors in $\mathbb{R}^m$ with $m > 1$, unlike points in $\mathbb{R}$, are not necessarily componentwise comparable; e.g., vector $(1, 2, 3)$ is neither "bigger," nor "smaller," nor equal to $(3, 2, 1)$. As an upshot, for an arbitrary u the optimal value function need not exist — no π's value function may dominate the values of all other policies everywhere.

2.3.1   EXPECTED LINEAR ADDITIVE UTILITY AND THE OPTIMALITY PRINCIPLE

Fortunately, there is a natural utility function that avoids both of the above pitfalls:

Definition 2.9   Expected Linear Additive Utility. An *expected linear additive utility* function is a function $u(R_t, R_{t+1}, \ldots) = \mathbb{E}[\sum_{t'=t}^{|\mathcal{D}|} \gamma^{t'-t} R_{t'}] = \mathbb{E}[\sum_{t'=0}^{|\mathcal{D}|-t} \gamma^{t'} R_{t'+t}]$ that computes the utility of a reward sequence as the expected sum of (possibly discounted) rewards in this sequence, where $\gamma \geq 0$ is the *discount factor*.

This definition of utility has an intuitive interpretation. It simply says that a policy is as good as the amount of discounted reward it is expected to yield. Setting $\gamma = 1$ expresses indifference of the agent to the time when a particular reward arrives. Setting it to a value $0 \leq \gamma < 1$ reflects various degrees of preference to rewards earned sooner. This is very useful, for instance, for modeling an agent's attitude to monetary rewards. The agent may value the money it gets today more than the same amount money it could get in a month, because today's money can be invested and yield extra income in a month's time.

Of course, expected linear additive utility is not always the best measure of a policy's quality. In particular, it assumes the agent to be *risk-neutral*, i.e., oblivious of the variance in the rewards yielded by a policy. As a concrete example of risk-neutral behavior, suppose the agent has a choice of either getting a million dollars or playing the following game. The agent would flip a fair coin, and if the coin comes up heads, the agent gets two million dollars; otherwise, the agent gets nothing. The two options can be interpreted as two policies, both yielding the expected reward of one million dollars. The expected linear additive utility model gives the agent no reason to prefer one policy over the other, since their expected payoff is the same. In reality, however, most people in such circumstances would tend to be *risk-averse* and select the option with lower variance, i.e., just take one million dollars. Although expected linear additive utility does not capture these nuances, it is still a convenient indicator of policy quality in many cases, and we will assume it in the rest of this book.

Why is expected linear additive utility so special? As it turns out, letting $V^\pi(h_{s,t}) = \mathbb{E}[\sum_{t'=0}^{|\mathcal{D}|-t} \gamma^{t'} R_{t'+t}^{\pi h_{s,t}}]$ guarantees a very important MDP property that, for now, we state informally as follows:

The Optimality Principle. *If every policy's quality can be measured by this policy's expected linear additive utility, there exists a policy that is optimal at every time step.*

The Optimality Principle is one of the most important results in the theory of MDPs. In effect, it says that if the expected utility of every policy is well-defined, then there is a policy that maximizes this utility in every situation, i.e., in every situation performs at least as well as any other policy.

The statement of the Optimality Principle has two subtle points. First, its claim is valid only if "every policy's quality can be measured by the policy's expected linear additive utility." When does this premise fail to hold? Imagine, for example, an MDP with an infinite $\mathcal{D}$, $\gamma = 1$, two states, s and s', and an action a s.t. $\mathcal{T}(s, a, s', t) = 1.0$, $\mathcal{T}(s', a, s, t) = 1.0$, $\mathcal{R}(s, a, s', t) = 1$, and $\mathcal{R}(s', a, s, t) = -1$ for all $t \in \mathcal{D}$. In other words, the agent can only travel in a loop between states s and s'. In this MDP, for both states the expected sum of rewards keeps oscillating (between 1 and 0 for s and between -1 and 0 for s'), never converging to any finite value or diverging to infinity. Second, and this is very important, *optimal policies are not necessarily unique.*

At the same time, as stated, the Optimality Principle may seem more informative than it actually is. Since we have not imposed any restrictions either on the γ parameter, on the sequence of the reward random variables $R_{t'+t}$, or on the number of time steps $|\mathcal{D}|$, the expectation $\mathbb{E}[\sum_{t'=0}^{|\mathcal{D}|-t} \gamma^{t'} R_{t'+t}]$ may be infinite. This may happen, for example, when the reward of any action in any state at any time step is at least $\epsilon > 0$, γ is at least 1, and the number of time steps is infinite. As a consequence, *all* policies in an MDP could be optimal under the expected linear additive utility criterion, since they all might have infinite values in all the states and hence would be indistinguishable from each other in terms of quality. The following sections will concentrate on ways to enforce the finiteness of any policy's value function, thereby ensuring that the optimality criterion is actually meaningful.

2.3.2   FINITE-HORIZON MDPS

Perhaps the easiest way to make sure that $\mathbb{E}[\sum_{t'=0}^{|\mathcal{D}|-t} \gamma^{t'} R_{t'+t}]$ is finite for any conceivable sequence of random reward variables in a given MDP is to limit the MDP to a finite number of time steps. In this case, the summation terminates after a finite number of terms $|\mathcal{D}| = T_{max}$, called the *horizon*, and the MDP is called a *finite-horizon* MDP.

Definition 2.10   Finite-Horizon MDP. A finite-horizon MDP is an MDP as described in Definition 2.1 with a finite number of time steps, i.e., with $|\mathcal{D}| = T_{max} < \infty$.

In most cases where finite-horizon MDPs are used, γ is set to 1, so the value of a policy becomes the expected total sum of rewards it yields. When analyzing algorithms for solving finite-horizon MDPs later in the book, we assume $\gamma = 1$, although other γ values do not change their properties.

One example [204] of a scenario appropriately modeled as a finite-horizon MDP is where the agent is trying to teach a set of skills to a student over T_{max} lessons (time steps). The agent can devote a lesson to teaching a new topic, giving a test to check the student's proficiency, or providing hints/additional explanations about a previously taught topic — this is the agent's action set. The agent gets rewarded if the student does well on the tests. The probability of a student doing well on them depends on her current proficiency level, which, in turn, depends on the amount of explanations and hints she has received from the agent. Thus, the agent's objective is to plan out the available lessons so as to teach the student well and have time to verify the student's knowledge via exams.

The Optimality Principle for finite-horizon MDPs can be restated in a precise form as follows [197].

Theorem 2.11   The Optimality Principle for Finite-Horizon MDPs. For a finite-horizon MDP with $|\mathcal{D}| = T_{max} < \infty$, define $V^\pi(h_{s,t}) = \mathbb{E}[\sum_{t'=0}^{T_{max}-t} R_{t'+t}^{\pi_{h_{s,t}}}]$ for all $1 \le t \le T_{max}$, and $V^\pi(h_{s,T_{max}+1}) = 0$. Then the optimal value function V^* for this MDP exists, is Markovian, and satisfies, for all $h \in \mathcal{H}$,

$$V^*(h) = \max_\pi V^\pi(h) \tag{2.1}$$

and, for all $s \in \mathcal{S}$ and $1 \le t \le T_{max}$,

$$V^*(s,t) = \max_{a \in \mathcal{A}} \left[\sum_{s' \in \mathcal{S}} \mathcal{T}(s,a,s',t)[\mathcal{R}(s,a,s',t) + V^*(s',t+1)] \right]. \tag{2.2}$$

Moreover, the optimal policy π^* corresponding to the optimal value function is deterministic Markovian and satisfies, for all $s \in \mathcal{S}$ and $1 \le t \le T_{max}$,

$$\pi^*(s,t) = \operatorname*{argmax}_{a \in \mathcal{A}} \left[\sum_{s' \in \mathcal{S}} \mathcal{T}(s,a,s',t)[\mathcal{R}(s,a,s',t) + V^*(s',t+1)] \right]. \tag{2.3}$$

This statement of the Optimality Principle is more concrete than the previous one, as it postulates the existence of *deterministic Markovian* optimal policies for finite-horizon MDPs. Their optimal value function dominates all other value functions, including general history-dependent ones, but satisfies the simpler Equation 2.2. Thus, as "promised" in Section 2.2, finite-horizon MDPs can be solved by restricting our attention to Markovian deterministic policies without sacrificing solution quality.

2.3.3   INFINITE-HORIZON DISCOUNTED-REWARD MDPS

Although finite-horizon MDPs have simple mathematical properties, this model is quite restrictive. In many scenarios, the reward is accumulated over an infinite (or virtually infinite) sequence of time steps. For instance, suppose we want to compute a policy for controlling the system of elevators in a building [204]. The elevators should work so as to minimize the average time between a person calling an elevator and arriving at their destination floor. This scenario has a nearly infinite horizon, since the elevators should continue to operate as long as the building exists. Therefore, we need to consider a way to ensure the convergence of an infinite weighted expected reward series.

Before we tackle this issue, however, we pause for a moment to describe how we can specify an infinite-horizon MDP concisely. Writing down this kind of MDP may be nontrivial for the same reason as describing a general MDP solution is. Recall from Section 2.2 that a non-stationary policy for an MDP with infinite $\mathcal{D}$ may take an infinite amount of space to enumerate, because such a policy depends on time. An MDP's transition function $\mathcal{T}$ and reward function $\mathcal{R}$ depend on time as well, and hence may be equally cumbersome. To cope with the potentially infinite policy description size, in Section 2.2 we introduced the notion of a stationary Markovian policy (Definition 2.5), which depends on state but not time and can always be characterized compactly as a result. In the same vein, we use stationarity to reduce the size of $\mathcal{T}$ and $\mathcal{R}$ when the horizon is infinite:

Definition 2.12   Stationary Transition Function. An MDP's transition function $\mathcal{T}$ is stationary if for any states $s_1, s_2 \in \mathcal{S}$ and action $a \in \mathcal{A}$, the value $\mathcal{T}(s_1, a, s_2, t)$ does not depend on t, i.e., $\mathcal{T}(s_1, a, s_2, t) = \mathcal{T}(s_1, a, s_2, t')$ for any $t, t' \in \mathcal{D}$.

Definition 2.13   Stationary Reward Function. An MDP's reward function $\mathcal{R}$ is stationary if for any states $s_1, s_2 \in \mathcal{S}$ and action $a \in \mathcal{A}$, the value $\mathcal{R}(s_1, a, s_2, t)$ does not depend on t, i.e., $\mathcal{R}(s_1, a, s_2, t) = \mathcal{R}(s_1, a, s_2, t')$ for any $t, t' \in \mathcal{D}$.

Since stationary $\mathcal{T}$ and $\mathcal{R}$ are constant with respect to $\mathcal{D}$, we will refer to them more concisely as $\mathcal{T} : \mathcal{S} \times \mathcal{A} \times \mathcal{S} \rightarrow [0, 1]$ and $\mathcal{R} : \mathcal{S} \times \mathcal{A} \times \mathcal{S} \rightarrow \mathbb{R}$. When the transition function is stationary, we will call s' an *outcome* of a in s whenever $\mathcal{T}(s, a, s') > 0$. We will call MDPs with stationary transition and reward functions *stationary MDPs*. All MDPs with infinite $\mathcal{D}$ that we will consider in this book are stationary.

Returning to the issue of an infinite reward series expectation, an easy way to force it to converge *for a stationary MDP* is to ensure that $\{\gamma^i\}_{i=0}^{\infty}$ forms a decreasing geometric sequence, i.e., to set γ to a value $0 \leq \gamma < 1$. Let us rewrite the general form of the utility function $u(R_t, R_{t+1}, \ldots) = \mathbb{E}[\sum_{t'=0}^{|\mathcal{D}|-t} \gamma^{t'} R_{t'+t}]$ as $\sum_{t'=0}^{\infty} \gamma^{t'} \mathbb{E}[R_{t'+t}]$. Since $\mathcal{R}$ is stationary and its domain $\mathcal{S} \times \mathcal{A} \times \mathcal{S}$ is finite, the terms $\mathbb{E}[R_{t'+t}]$ cannot grow without bound, i.e., for all $t' \geq 0$, $\mathbb{E}[R_{t'+t}] \leq M$ for some finite $M \in \mathbb{R}$. Therefore, $u(R_t, R_{t+1}, \ldots) \leq \sum_{t'=0}^{\infty} \gamma^{t'} M = \frac{M}{1-\gamma}$. Thus, defining $V^{\pi}(h_{s,t}) = u(R_t^{\pi_{h_{s,t}}}, R_{t+1}^{\pi_{h_{s,t}}}, \ldots) = \mathbb{E}[\sum_{t'=0}^{\infty} \gamma^{t'} R_{t'+t}]$ for $0 \leq \gamma < 1$ gives us a value function that is finite for all policies π in all situations.

Definition 2.14   Infinite-Horizon Discounted-Reward MDP. An infinite-horizon discounted-reward MDP is a stationary MDP as described in Definition 2.1, in which the value of a policy is defined as $V^{\pi}(h_{s,t}) = \mathbb{E}[\sum_{t'=0}^{\infty} \gamma^{t'} R_{t'+t}^{\pi_{h_{s,t}}}]$, where the discount factor γ is a model parameter restricted to be $0 \leq \gamma < 1$.

Besides having the pleasing mathematical property of a bounded value function, discounted-reward MDPs also have a straightforward interpretation. They model problems in which the agent gravitates toward policies yielding large rewards in the near future rather than policies yielding similar rewards but in a more distant future. At the same time, any particular choice of γ is usually hard to justify. One exception is the field of finance, where γ can model the rate of inflation or the interest rate, readily available from economic data.

As in the case with finite-horizon MDPs, the Optimality Principle can be specialized to infinite-horizon discounted-reward MDPs, but in an even stronger form due to the assumptions of stationarity [197].

Theorem 2.15   The Optimality Principle for Infinite-Horizon MDPs. For an infinite-horizon discounted-reward MDP with discount factor γ s.t. $0 \leq \gamma < 1$, define $V^{\pi}(h_{s,t}) = \mathbb{E}[\sum_{t'=0}^{\infty} \gamma^{t'} R_{t'+t}^{\pi_{h_{s,t}}}]$. Then the optimal value function V^* for this MDP exists, is stationary Markovian, and satisfies, for all $h \in \mathcal{H}$,

$$V^*(h) = \max_{\pi} V^{\pi}(h) \tag{2.4}$$

and, for all $s \in \mathcal{S}$,

$$V^*(s) = \max_{a \in \mathcal{A}} \left[\sum_{s' \in \mathcal{S}} \mathcal{T}(s, a, s')[\mathcal{R}(s, a, s') + \gamma V^*(s')] \right]. \tag{2.5}$$

Moreover, the optimal policy π^* corresponding to the optimal value function is deterministic stationary Markovian and satisfies, for all $s \in \mathcal{S}$,

$$\pi^*(s) = \operatorname*{argmax}_{a \in \mathcal{A}} \left[\sum_{s' \in \mathcal{S}} \mathcal{T}(s, a, s')[\mathcal{R}(s, a, s') + \gamma V^*(s')] \right]. \tag{2.6}$$

2.3.4 INDEFINITE-HORIZON MDPS

Finite-horizon and infinite-horizon discounted MDPs are located at the opposite ends of a spectrum — the former assume the number of time steps to be limited and known; the latter assume it to be infinite. This suggests a natural third possibility, problems with finite but unknown horizon.

Imagine, for instance, a space shuttle trying to enter Earth's atmosphere and land after a mission. During every revolution around Earth, the shuttle can either choose to land or continue on to the next revolution, e.g., due to bad weather. In theory, the shuttle can orbit in space forever. However, poor weather conditions cannot persist for too long, so with overwhelming probability the shuttle will land after some finite but *a priori* unknown number of attempts, if controlled optimally. More formally, we can think of indefinite-horizon problems as processes that, under every optimal policy, inevitably reach some terminal state. As long as the process has not reached a terminal state, it incurs negative reward, i.e., cost (in the shuttle example, the cost can be crew fatigue). Once the target state is reached, no more cost is incurred. Thus, the total expected cost of any optimal policy in such an MDP should be finite even without a discount factor.

In spite of this fairly straightforward logic, precisely stating a set of conditions under which an optimal policy exists in the presence of an indefinite horizon is tricky, and we devote the entire next section to its derivation.

2.4 STOCHASTIC SHORTEST-PATH MDPS

In the previous section, we introduced indefinite-horizon MDPs, a class of problems in which every optimal policy eventually leads the system to some terminal state. However, the informal definition we gave for it is somewhat unsatisfying, as it lets us recognize an indefinite-horizon MDP only once we solve it. A more useful definition would instead specify more basic restrictions on an MDP to *a priori* guarantee the existence of an indefinite-horizon optimal policy.

2.4.1 DEFINITION

We now turn to formulating a set of conditions under which optimal solutions of an MDP are policies with indefinite horizon. Doing so will let us derive a class of problems that is widely studied in AI, the *stochastic shortest-path MDPs*. To start with, we formalize the intuition behind the concept of a policy that eventually reaches a terminal state, as described in Section 2.3.4. We do so by introducing the definition of a *proper policy*, a policy that is guaranteed to bring the agent to a goal state from any other state.

Definition 2.16 **Proper policy.** For a given stationary MDP, let $\mathcal{G} \subseteq \mathcal{S}$ be a set of *goal* states s.t. for each $s_g \in \mathcal{G}$ and all $a \in \mathcal{A}$, $\mathcal{T}(s_g, a, s_g) = 1$ and $\mathcal{R}(s_g, a, s_g) = 0$. Let h_s be an execution history that terminates at state s. For a given set $S' \subseteq \mathcal{S}$, let $P_t^\pi(h_s, S')$ be the probability that after execution history h_s, the agent transitions to some state in S' within t time steps if it follows policy π. A policy π is called *proper at state s* if $P_t^\pi(h_s, \mathcal{G}) = 1$ for all histories $h_s \in \mathcal{H}$ that terminate at s.

If for some h_s, $\lim_{t \to \infty} P_t^\pi(h_s, \mathcal{G}) < 1$, π is called *improper at state s*. A policy π is called *proper* if it is proper at all states $s \in \mathcal{S}$. Otherwise, it is called *improper*.

Without loss of generality, we can assume that there is only one goal state — if there are more, we can add a special state with actions leading to it from all the goal states and thus make this state the de-facto unique goal. Therefore, as a shorthand, in the rest of the book we will sometimes say that a policy reaches *the goal*, meaning that it reaches some state in $\mathcal{G}$. We will say that a policy reaches the goal with probability 1 about a policy that is guaranteed to eventually reach some state in $\mathcal{G}$, although possibly not the same state in every execution.

As Definition 2.16 implies, a proper policy is, in general, indefinite-horizon. Thus, all we need to do to ensure that an MDP is an infinite-horizon MDP is to impose conditions on it guaranteeing that at least one of its optimal policies is proper. One way to do so when designing an MDP is as follows. Suppose we make the rewards for all actions in all states of our MDP negative. The agent will effectively have to pay a cost for executing every action. Suppose also that our MDP has at least one proper policy. Under such conditions, the longer an agent stays away from the goal state, the more cost the agent is likely to pay. In particular, if the agent uses an improper policy, it will never be able to reach the goal from some states, i.e., the total expected cost of that policy from some states will be infinite. Contrariwise, for every proper policy we expect this cost to be finite for every state. As an upshot, some proper policy will be preferable to all improper ones and at least as good as any other proper policy, which is exactly what we want. A kind of indefinite-horizon MDP called *stochastic shortest-path MDP* captures this intuition.

Definition 2.17   Stochastic Shortest-Path MDP. *(Weak definition.)* A *stochastic shortest-path (SSP) MDP* is a tuple $\langle \mathcal{S}, \mathcal{A}, \mathcal{T}, \mathcal{C}, \mathcal{G} \rangle$ where:

- $\mathcal{S}$ is the finite set of all possible states of the system,

- $\mathcal{A}$ is the finite set of all actions an agent can take,

- $\mathcal{T} : \mathcal{S} \times \mathcal{A} \times \mathcal{S} \to [0, 1]$ is a stationary transition function specifying the probability $\mathcal{T}(s_1, a, s_2)$ of going to state s_2 whenever action a is executed when the system is in state s_1,

- $\mathcal{C} : \mathcal{S} \times \mathcal{A} \times \mathcal{S} \to [0, \infty)$ is a stationary cost function that gives a finite strictly positive cost $\mathcal{C}(s_1, a, s_2) > 0$ incurred whenever the system goes from state s_1 to state s_2 as a result of executing action a, with the exception of transitions from a goal state, and

- $\mathcal{G} \subseteq \mathcal{S}$ is the set of all goal states s.t. for every $s_g \in \mathcal{G}$, for all $a \in \mathcal{A}$, and for all $s' \notin \mathcal{G}$, the transition and cost functions obey $\mathcal{T}(s_g, a, s_g) = 1$, $\mathcal{T}(s_g, a, s') = 0$, and $\mathcal{C}(s_g, a, s_g) = 0$,

under the following condition:

- There exists at least one proper policy.

Although superficially similar to the general MDP characterization (Definition 2.1), this definition has notable differences. Other than the standard stationarity requirements, note that instead of a reward function SSP MDPs have a cost function $\mathcal{C}$, since the notion of cost is more appropriate for scenarios modeled by SSP MDPs. Consequently, as we discuss shortly, solving an SSP MDP means finding an *expected-cost minimizing*, as opposed to a reward-maximizing, policy. Further, the set of decision epochs $\mathcal{D}$ has been omitted, since for every SSP MDP it is implicitly the set of all natural numbers. These differences are purely syntactic, as we can define $\mathcal{R} = -\mathcal{C}$ and go back to the reward-maximization-based formulation. A more fundamental distinction is the presence of a special set of (terminal) goal states, staying in which forever incurs no cost. Last but not least, the SSP definition requires that SSP MDPs have at least one policy that reaches the goal with probability 1.

The Optimality Principle for SSP MDPs [21] holds in a slightly modified form that nonetheless postulates the existence of simple optimal policies, as for the infinite-horizon discounted-cost MDPs.

Theorem 2.18   The Optimality Principle for Stochastic Shortest-Path MDPs. For an SSP MDP, define $V^{\pi}(h_{s,t}) = \mathbb{E}[\sum_{t'=0}^{\infty} R_{t'+t}^{\pi_{h_{s,t}}}]$. Then the optimal value function V^* for this MDP exists, is stationary Markovian, and satisfies, for all $h \in \mathcal{H}$,

$$V^*(h) = \min_{\pi} V^{\pi}(h) \qquad (2.7)$$

and, for all $s \in \mathcal{S}$,

$$V^*(s) = \min_{a \in \mathcal{A}} \left[\sum_{s' \in \mathcal{S}} \mathcal{T}(s, a, s')[\mathcal{C}(s, a, s') + V^*(s')] \right]. \qquad (2.8)$$

Moreover, the optimal policy π^* corresponding to the optimal value function is deterministic stationary Markovian and satisfies, for all $s \in \mathcal{S}$,

$$\pi^*(s) = \operatorname*{argmin}_{a \in \mathcal{A}} \left[\sum_{s' \in \mathcal{S}} \mathcal{T}(s, a, s')[\mathcal{C}(s, a, s') + V^*(s')] \right]. \qquad (2.9)$$

The largest differences from the previous versions of this principle are the replacement of maximizations with minimizations, the replacement of a reward function with a cost function, and the omitted γ factor, since it is fixed at 1. We emphasize once again that the practical importance of the existence of deterministic stationary Markovian policies as optimal solutions is hard to overestimate. It guarantees that an optimal solution for many interesting MDPs is just a mapping from states to actions. This greatly simplifies designing efficient algorithms for solving such models. To get an idea of the efficiency gains the existence of optimal deterministic stationary Markovian policies might give us, consider a brute-force algorithm for solving MDPs. By definition, it would try to go through all possible policies (solutions), evaluate each of them somehow, and choose the best one. Even assuming a finite-time algorithm for evaluating a policy exists, without the result above guaranteeing the existence of an optimal deterministic stationary Markovian policy, the brute-force strategy would never halt, since the number of policies of the form in Definition 2.3 is generally infinite. However, thanks to the Optimality Principle, we only need to enumerate all the deterministic stationary Markovian policies to find an optimal one. There are "merely" $|\mathcal{A}|^{|\mathcal{S}|}$ of them, so our exhaustive search would terminate in a finite (although possibly large) amount of time. In the rest of the book, we will use the term "policy" to refer to deterministic stationary Markovian policies, unless explicitly stated otherwise.

The following theorem, a corollary of several results from the SSP MDP theory [21], sheds more light on the semantics of this model.

Theorem 2.19   In an SSP MDP, $V^*(s)$ is the smallest expected cost of getting from state s to the goal. Every optimal stationary deterministic Markovian policy in an SSP MDP is proper and minimizes the expected cost of getting to the goal at every state.

This theorem explains why the MDPs we are discussing are called stochastic shortest-path problems. It says that, under the SSP MDP definition conditions, policies that optimize for the expected undiscounted linear additive utility also optimize for the expected cost of reaching the goal. In other words, in expectation they are the "shortest" ways of reaching the goal from any given state. To see this analogy in a different way, consider an SSP MDP with a single goal state and a deterministic transition function, under which a given action a used in state s always leads the system into the same state s'. What is the resulting problem reminiscent of? Indeed, it is the classical problem of finding a weighted *shortest path in a graph*.

In the AI literature (e.g., [21]), one also encounters another, broader definition of SSP MDPs:

Definition 2.20   **Stochastic Shortest-Path MDP.** *(Strong definition.)* A stochastic shortest-path (SSP) MDP is a tuple $\langle \mathcal{S}, \mathcal{A}, \mathcal{T}, \mathcal{C}, \mathcal{G} \rangle$ where $\mathcal{S}, \mathcal{A}, \mathcal{T} : \mathcal{S} \times \mathcal{A} \times \mathcal{S} \to [0, 1], \mathcal{G} \subseteq \mathcal{S}$ are as in Definition 2.17, and $\mathcal{C} : \mathcal{S} \times \mathcal{A} \times \mathcal{S} \to \mathbb{R}$ is a stationary *real-valued* cost function, under two conditions:

- There exists at least one proper policy,

- For every improper stationary deterministic Markovian policy π, for every $s \in \mathcal{S}$ where π is improper, $V^\pi(s) = \infty$.

Theorem 2.19 holds for this SSP definition as well as for the weak one (Definition 2.17), so optimal solutions to SSP MDPs defined this way still include the stationary deterministic Markovian policies that minimize the expected cost of getting to the goal at every state.

The strong definition can be shown to be more general than the weak one. The former does not impose any local constraints on the cost function — its value can be any real number, positive or negative. However, it adds an additional constraint insisting that every improper policy has an infinite value in at least one state. The net effect of these requirements is the same; in both cases *some* proper policy for the SSP MDP is preferable to all improper ones. Thus, instead of imposing local constraints on the cost values, Definition 2.20 introduces a global one. On the positive side, this allows more choices for the cost function, and hence admits more MDPs than the weak definition. On the other hand, the weak definition is more intuitive and, given an MDP, provides an easier way to determine whether the MDP is an SSP problem or not. In the rest of the book, when we discuss SSP MDPs, we will be implicitly using their strong, more general definition.

2.4.2   STOCHASTIC SHORTEST-PATH MDPS AND OTHER MDP CLASSES

Our discussion from section 2.3 up to this point has presented three separate MDP classes for which an optimal solution is both well-defined and guaranteed to exist — finite-horizon, infinite-horizon discounted-reward, and SSP MDPs. As we show in this subsection, this view is not entirely accurate. SSP MDPs are, in fact, strictly more general than the other MDP classes and properly contain them, as the following theorem states.

Theorem 2.21   Denote the set of all finite-horizon MDPs as *FH*, the set of all infinite-horizon discounted-reward MDPs as *IFH*, and the set of all SSP MDPs as *SSP*. The following statements hold true [24]:

$$FH \subset SSP$$

$$IFH \subset SSP$$

We do not provide a formal proof of this fact, but outline its main idea. Our high-level approach will be to show that each finite-horizon and infinite-horizon discounted-reward MDP can be compiled into an equivalent SSP MDP. Let us start with finite-horizon MDPs. Suppose we are given an FH MDP $\langle S, A, D, T, R \rangle$. Replace the state space S of such an MDP with the set $S' = S \times D$. This transformation makes the current decision epoch part of the state and is the key step in our proof. Replace T with a stationary transition function T' s.t. $T'((s, t), a, (s', t + 1)) = p$ whenever $T(s, a, s') = p$. By similarly changing R, construct a new reward function R' and replace R with a cost function $C = -R'$. Finally, assuming $|D| = T_{max} < \infty$, let the set of goal states be $G = \{(s, T_{max}) \mid s \in S\}$. The tuple we just constructed, $\langle S', A, T', C, G \rangle$, clearly satisfies the strong definition of SSP MDP.

Now suppose we are given an IFH MDP $\langle S, A, D, T, R \rangle$ with a discount factor γ. The main insight in transforming this MDP into an SSP problem is to replace R with $C = -R$ and to add a special goal state s_g to S, where the system can transition at any decision epoch with probability γ

using any action from any state (the transition function $\mathcal{T}$ needs to be scaled so that the probabilities of all other outcomes sum to $1 - \gamma$). Transitioning to s_g from any state using any action will incur the cost of 0. It is easy to verify that the new MDP conforms to the strong SSP MDP definition with $\mathcal{G} = \{s_g\}$, and it can be shown that every stationary deterministic Markovian policy in it is optimal if and only if it is optimal in the original IFH MDP.

Theorem 2.21's significance is in allowing us to concentrate on developing algorithms for *SSP*, since *FH* and *IFH* are merely its subclasses. Of course, since *FH* and *IFH* are more specialized, some techniques that do not extend to *SSP* can handle them more efficiently than the algorithms applying to *SSP* as a whole. While most techniques in this book target *SSP* as the more general class, we will also occasionally discuss important algorithms that work only for *FH* and *IFH*.

2.5   FACTORED MDPS

Although defining an MDP class takes just a few lines of text, *describing an MDP instance* can be much more cumbersome. The definitions of various MDP classes tell us that an MDP should have a set of states, a set of actions, and other elements, but how do we specify them to an MDP solver? For example, take the state space. One way to describe it is to assign a distinct number to every state of the system, i.e., let $\mathcal{S} = \{1, 2, 3, \cdots\}$. Such a state space description is called *atomic* or *flat*. It is inconvenient for at least two reasons. First, the set of states in a scenario we may want to model can be extremely large, bigger than the number of elementary particles in the universe. Enumerating such sets is impractical; it would take an MDP solver centuries of CPU time just to read the problem description. Second, this description is very uninformative for an MDP solver. It does not convey the structure of the particular problem at hand, which could otherwise let the solver perform some optimizations. These arguments apply to flat representations of other parts of the MDP, such as the transition function, as well. In fact, describing a stationary transition function in this way would take significantly more space and time than describing the state space itself, $O(|\mathcal{S}|^2|\mathcal{A}|)$.

2.5.1   FACTORED STOCHASTIC SHORTEST-PATH MDPS

To avoid the drawbacks of a flat representation, one can instead specify each state as a combination of values of several *state variables* relevant to the problem at hand, i.e., *factor* the state space into constituent variables. Doing so helps compactly describe both the state space and other MDP components, as shown in the following definition of *factored SSP MDPs* and illustrated with examples afterwards:

Definition 2.22   Factored SSP MDP. A *factored SSP MDP* is a tuple $\langle \mathcal{X}, \mathcal{S}, \mathcal{A}, \mathcal{T}, \mathcal{C}, \mathcal{G} \rangle$ obeying the conditions in Definition 2.20, where

- $\mathcal{X} = \{X_1, \ldots, X_n\}$ is a set of *state variables* (sometimes also called *features* or *domain variables*) whose domains are sets $dom(X_1), \ldots, dom(X_n)$ respectively,

- The finite state space is represented as a set $\mathcal{S} = dom(X_1) \times \ldots \times dom(X_n)$,

- The finite set of all actions is $\mathcal{A}$,

- The transition function is represented as a mapping $\mathcal{T} : (dom(X_1) \times \ldots \times dom(X_n)) \times \mathcal{A} \times (dom(X_1) \times \ldots \times dom(X_n)) \to [0, 1]$

- The cost function is represented as a mapping $\mathcal{C} : (dom(X_1) \times \ldots \times dom(X_n)) \times \mathcal{A} \times (dom(X_1) \times \ldots \times dom(X_n)) \to \mathbb{R}$, and

- The goal set $\mathcal{G}$ is represented as a set of states satisfying a specified logical formula over assignments of values to variables in $\mathcal{X}$.

This definition states what a factored MDP is at an abstract level, without specifying how exactly each of its components, e.g., the transition function, should be described. In what follows we examine one way of formally specifying factored MDP formally and compactly. Before doing this, however, we introduce some additional terminology. We call an assignment of a value $x_i \in dom(X_i)$ to a variable X_i a *literal over* X_i. For simplicity of discussion, in this book we will assume the state variables to be binary, i.e., $dom(X_i) = \{True, False\}$ for all i. All finite discrete factored MDPs can be converted into this binary form. We call a literal that assigns the *True* value to its variable X_i a *positive* literal and denote it, with a slight abuse of notation, X_i, just like the variable itself. Similarly, we call a literal that assigns the *False* value a *negative* literal and denote it as $\neg X_i$.

2.5.2   PPDDL-STYLE REPRESENTATION

According to the first factored representation we examine, an MDP's components have the following form:

- The set $\mathcal{X}$ is given as a collection of appropriately named variables with specified domains. For a binary factored MDP, it may appear as follows: $\mathcal{X} = \{A : \{True, False\}, B : \{True, False\}, C : \{True, False\}, D : \{True, False\}\}$.

- The state space $\mathcal{S}$ is the set of all conjunctions of literals over all the state variables. Again assuming binary variables, an example of a state is $s = A \wedge \neg B \wedge \neg C \wedge D$.

- Each action in the set $\mathcal{A}$ is a tuple of the form

$$\langle prec, \langle p_1, add_1, del_1 \rangle, \cdots, \langle p_m, add_m, del_m \rangle \rangle$$

s.t. all such tuples in $\mathcal{A}$ together specify the transition function $\mathcal{T}(s, a, s')$ as follows:

  - *prec* is the action's *precondition* represented as a conjunction of literals. We interpret an action's precondition to say that the action may cause a transition only from those states that have all of the literals in the precondition. More formally, we say that the action is *applicable* only in states where the precondition *holds*. The consequences of applying an action in states where it is not applicable are undefined, i.e., $\mathcal{T}(s, a, s')$ is undefined for all triples s, a, s' where a's precondition does not hold in s. Notice that this contradicts Definition 2.20, which requires $\mathcal{T}(s, a, s')$ to be defined for all triples s, a, s'. However, we overlook this technicality, since it does not invalidate any of the SSP MDP results considered in this book.

- $\langle p_1, add_1, del_1 \rangle, \cdots, \langle p_m, add_m, del_m \rangle$ is the list of the action's *probabilistic effects*. The description of an effect consists of the effect's probability p_i and two lists of changes the effect makes to the state. In particular, add_i, the *i-th add effect*, is the conjunction of positive literals that the i-th effect adds to the state description. Similarly, the *i-th delete effect* is the conjunction of negative literals that the i-th effect adds to the state description. Implicitly, inserting a positive literal removes the negative literal over the same variable, and vice versa. For each pair of an action's probabilistic effects, we assume that the effects are distinct, i.e., that either $add_i \neq add_j$ or $del_i \neq del_j$ or both.

We interpret this specification to say that if an action is applied in a state where its precondition holds, nature will "roll the dice" and select one of the action's effects according to the effects' probabilities. The chosen effect will cause the system to transition to a new state by making changes to the current state as dictated by the corresponding add and delete effect. Formally, $\mathcal{T}(s, a, s') = p_i$ whenever a is applicable in s and s' is the result of changing s with the i-th effect.

Example: Suppose we apply action $a = \langle A, \langle 0.3, B \wedge C, \bigwedge \emptyset \rangle, \langle 0.7, B, \neg A \wedge \neg D \rangle \rangle$ in state $s = A \wedge \neg B \wedge \neg C \wedge D$, where $\bigwedge \emptyset$ stands for the empty conjunction. The action's precondition, the singleton conjunction A, holds in s, so the action is applicable in this state. Suppose that nature chooses effect $\langle 0.7, B, \neg A \wedge \neg D \rangle$. This effect causes the system to transition to state $s' = \neg A \wedge B \wedge \neg C \wedge \neg D$. □

In the rest of the book, we will refer to the precondition of a as $prec(a)$, to the i-th add effect of a as $add_i(a)$, and to the i-th delete effect of a as $del_i(a)$.

- We assume the description of the cost function $\mathcal{C}(s, a, s')$ to be polynomial in $|\mathcal{X}|$, the number of state variables, and $|\mathcal{A}|$. As an example of such a specification, we could associate a *cost* attribute with each action $a \in \mathcal{A}$ and set $cost(a) = c_a$. The *cost* attribute's value c_a should be interpreted as the cost incurred by using a in any state s where a is applicable, independently of the state s' where the agent transitions as a result of applying a in s. We let $cost(a)$ (and hence $\mathcal{C}(s, a, s')$) be undefined whenever a is not applicable in s. Again, this clause violates Definition 2.20 but all the important SSP MDP results hold in spite of it.

- Finally, we assume the goal set $\mathcal{G}$ to be implicitly given by a conjunction of literals over a subset of the variables in $\mathcal{X}$ that any goal state has to satisfy. For instance, if the goal-describing conjunction is $A \wedge \neg C \wedge D$, then there are two goal states in our MDP, $A \wedge B \wedge \neg C \wedge D$ and $A \wedge \neg B \wedge \neg C \wedge D$.

The formalism we just described is very general. There is at least one MDP description language, PPDDL (Probabilistic Planning Domain Definition Language) [252], that expresses factored MDPs in this way, and in this book will call MDP formulations as above *PPDDL-style*. A PPDDL-style description is convenient when actions change variables in a correlated manner, i.e., have *correlated effects* and have just a few outcomes. For instance, under the aforementioned action $a = \langle A, \langle 0.3, B \wedge C, \bigwedge \emptyset \rangle, \langle 0.7, B, \neg A \wedge \neg D \rangle \rangle$, the sets of variables $\{B, C\}$ and $\{A, B, D\}$ always change their values together, and a has only two outcomes.

2.5.3   RDDL-STYLE REPRESENTATION

At the same time, for some MDPs, a PPDDL-style description may not be the most convenient choice. E.g., consider the problem of managing a network of n servers, in which every running server has some probability of going down and every server that is down has some probability of restarting at every time step if the network administrator does not interfere. We will refer to this scenario as *Sysadmin* throughout the book. In a formal factored specification of Sysadmin, its state space could be given by a set of binary state variables $\{X_1, \ldots, X_n\}$, each variable indicating the status of the corresponding server. Thus, the state space size would be 2^n. We would have a *noop* action that applies in every state and determines what happens to each server at the next time step if the administrator does nothing. Note, however, that, since each server can go up or down independently from others (i.e., variables change values in an *uncorrelated* fashion), at each time step the system can transition from the current state to any of the 2^n states. Therefore, to write down a description of *noop*, we would need to specify 2^n outcomes — an enormous number even for small values of n. In general, such actions are common in scenarios where many objects evolve independently and simultaneously, as in Sysadmin. They also arise naturally when the MDP involves *exogenous events*, changes that the agent cannot control and that occur in parallel with those caused by the agent's own actions. Examples of exogenous events include natural cataclysms and, sometimes, actions of other agents.

An MDP like this is more compactly formulated as a Dynamic Bayesian Network (DBN) [71], a representation particularly well suited for describing actions with uncorrelated effects. The most common MDP description language that expresses an MDP as a DBN is RDDL (Relational Dynamic Influence Diagram Language) [204], so we will call this representation *RDDL-style*. Under this representation, for each action and each domain variable, an engineer needs to specify a conditional probability distribution (CPD) over the values of that variable in the next state if the action is executed. Suppose, for instance, we want to say that if the status of the i-th server is "up" at time step t and the administrator does not intervene, then with probability 0.5 it will remain "up" at the next time step; similarly, if it is currently "down," it will remain "down" with probability 0.9. Then the CPD $P(X_i^{t+1} = \text{"up"}|X_i^t = \text{"up"}, Action = noop) = 0.5$, $P(X_i^{t+1} = \text{"down"}|X_i^t = \text{"down"}, Action = noop) = 0.9$ fully states the *noop*'s effect on variable X_i.

The advantage of an RDDL-style representation is that the size of each action's description, as illustrated by the above example, can be only polynomial in the number of domain variables when in the case of a PPDDL-style representation it would be exponential. At the same time, expressing *correlated* effects with DBNs can be tedious. Thus, the PPDDL-style and RDDL-style representations are largely complementary, and the choice between them depends on the problem at hand.

2.5.4 FACTORED REPRESENTATIONS AND SOLVING MDPS

Making MDPs easier to specify is only one motivation for using a factored representation. Another, and perhaps a more significant one, is increasing the efficiency of methods for finding and representing MDP solutions.

Notice that if S is specified in a flat form, all functions whose domain is S, such as value functions and policies, are effectively forced to be in a tabular form. In other words, they are forced to express mappings involving S explicitly, as a set of pairs, as opposed to implicitly, as an output of mathematical operations on members of S. To make this observation more concrete, consider a flat MDP with four states, whose optimal value function is $V^*(s_1) = 0$, $V^*(s_2) = 1$, $V^*(s_3) = 1$, $V^*(s_4) = 2$. Now, suppose this MDP can actually be meaningfully cast into a factored form as a problem with two binary state variables, X_1 and X_2. Let us denote the *True* value of a variable as 1 and *False* as 0. This MDP's optimal value function becomes $V^*((0, 0)) = 0$, $V^*((0, 1)) = 1$, $V^*((1, 0)) = 1$, $V^*((1, 1)) = 2$. A brief inspection reveals that V^* can be described implicitly with a simple formula, $V^*((X_1, X_2)) = X_1 + X_2$. The implicit formulaic specification is clearly more compact than the explicit tabular one. The difference is even more noticeable if we imagine a similar MDP with n state variables, and hence 2^n states. The size of its explicitly described optimal value function grows linearly with the size of its state space, while the size of the implicitly specified one, enabled by a factored representation, grows only logarithmically in it.

Thus, factored representations allow us to store value functions and policies more compactly than is possible with a flat specification. This applies not only to the optimal value functions and policies, but also to arbitrary ones. Throughout the book, we will see several approaches that exploit factored representations in order to make MDP solvers faster and less memory-hungry, including dimensionality reduction techniques (Section 6.4) and symbolic algorithms (Chapter 5). For now, we just mention that the increase in efficiency due to the use of factored representations can be dramatic not only in theory, but in practice as well.

2.6 COMPLEXITY OF SOLVING MDPS

Before delving into techniques for solving MDPs, we examine their computational complexity to get an idea of the efficiency we can hope these techniques to have. When analyzing the computational complexity of a problem class, one needs to keep in mind that the complexity is defined in terms of the input size. For MDPs, the input size is the size of an MDP's description. We have seen two of ways of describing MDPs:

- By specifying them in a flat representation, i.e., by explicitly enumerating their state space, action space, etc., and

- By specifying them in a factored representation, i.e., in terms of a set of variables $X_1, \ldots, X_n$.

The key observation for the MDP complexity analysis is that for a given factored MDP, its explicitly enumerated state space is exponential in the number of state variables — if the number of state variables is $|\mathcal{X}|$, then its state space size is $2^{|\mathcal{X}|}$. Similarly, flat descriptions of the transition and reward/cost functions are exponential in the size of their factored counterparts. As we have already observed, a factored representation can make an MDP much easier to specify. However, as the results below show, it makes them look difficult to solve with respect to their input size.

We begin by characterizing the complexity of solving MDPs in the flat representation.

Theorem 2.23   Solving finite-horizon MDPs in the flat representation is P-hard. Solving infinite-horizon discounted-reward and SSP MDPs in the flat representation is P-complete [191].

Notice that finite-horizon MDPs are not known to be in P, because solving them requires specifying an action for each *augmented state* in the set $\mathcal{S} \times \mathcal{D}$, which, in turn, could be exponential in the size of S if $|\mathcal{D}| = 2^{|S|}$. The above result also indicates that even infinite-horizon and SSP MDPs are some of the hardest polynomially solvable problems. In particular, they likely cannot benefit significantly from parallelization [191].

Since casting MDPs into a factored representation drastically reduces their description size, factored MDPs belong to a different computational complexity class.

Theorem 2.24   Factored finite-horizon, infinite-horizon discounted-reward, and SSP MDPs are *EXPTIME*-complete [96; 156].

EXPTIME-complete problem classes are much harder than P-complete or P-hard ones — the former are known to contain problems not solvable in polynomial time on modern computer architectures. However, it is not that factored representation makes solving MDPs very difficult; rather, it makes specifying them very easy compared to the hardness of solving them.

Factored MDPs' computational complexity can be somewhat reduced by imposing reasonable restrictions on their properties. In many situations, we are interested in finding a *partial* policy that reaches the goal (recall that finite-horizon and infinite-horizon discounted-reward MDPs can be converted into the goal-oriented form) *from a given state*, as opposed to from every state of the MDP. MDPs modeling such scenarios are said to have an *initial state*; we discuss them in more detail in Section 4.1. As an example, consider again the game of blackjack; in a given game, we are interested in how to win starting with the specific hand that we are dealt in that game, not from any hand we theoretically could have been dealt.

When the initial state is known, we can make the additional assumption that optimal policies that reach the goal from the initial state do so in a maximum number of steps polynomial in the number of state variables. Intuitively, this amounts to assuming that if we start executing the policy at the initial state, we will visit only a "small" number of states before ending up at the goal. This assumption often holds true in practice; indeed, non-contrived problems requiring policies that visit a number of states exponential in the number of state variables (i.e., linear in the size of the state space) are not so common. Moreover, executing such a policy would likely take a very long time, making the use of such a policy impractical even if we could obtain it reasonably quickly. To get a sense for how long an execution of such a policy would take, observe that in an MDP with only 100 binary state variables such a policy would need to visit on the order of 2^{100} states.

For MDPs with an initial state and a polynomially sized optimal policy, the following result in Theorem 2.25 holds.

Theorem 2.25   Factored finite-horizon, infinite-horizon discounted-reward, and SSP MDPs with an initial state in which an optimal policy reaches the goal from the initial state in a maximum number of steps polynomial in the number of state variables are *PSPACE*-complete [96; 156].

Importantly, with or without these assumptions, factored MDPs are hard not just to solve optimally but even to approximate, with no hope of discovering more efficient methods in the future unless a series of equivalences such as $P = NP$ are proven to hold. Since factored representations are dominant in AI, the above results have pushed the MDP research in the direction of methods that improve the efficiency of solving MDPs in practice, not in theory. Starting from the next chapter, we will examine many algorithms that achieve this.

CHAPTER 3

Fundamental Algorithms

We begin our journey of solving MDPs by first focusing on a fundamental set of techniques that forms the basis for most of the advanced algorithms. As in much of the rest of the book, this chapter focuses on MDPs with finite state and action spaces. We describe all algorithms and theoretical results using the Stochastic Shortest Path (SSP) formalism from Section 2.4.

Recall that an SSP MDP is a relatively more general MDP model, since it encompasses several other common MDP classes. The algorithms in this chapter apply to all SSP subclasses. However, some subclasses are amenable to tighter theoretical analyses and different algorithms. Sections 3.8-3.9 report some of the main results that apply only to specific subclasses of SSP MDPs.

We start by designing algorithms that compute the *best* policy, i.e., for each state, the best action that an agent should execute in that state. We will describe algorithms that compute stationary deterministic Markovian policies (i.e., $\pi : \mathcal{S} \to \mathcal{A}$). This is sufficient, since the optimality principle from Theorem 2.18 asserts that each SSP MDP has an optimal stationary deterministic Markovian policy π^*. Hereafter, any use of the term 'policy' denotes a deterministic mapping from states to actions.

Our objective in this chapter is to give an overview of *optimal* algorithms that produce *complete* policies, i.e., compute the solution for the entire state space:

Definition 3.1   Complete Policy. A policy is *complete* if its domain is all of the state space ($\mathcal{S}$).

Recall that the objective of an SSP MDP is to reach one of the possibly many absorbing goal states ($\mathcal{G}$) while minimizing the expected cost. Notice that a domain may have several best actions for a state, since they may all minimize the expected cost to a goal. Therefore, an MDP may have several optimal policies. Our aim is to design algorithms that compute any one of the possibly many optimal policies.

We remind the reader about two key assumptions in the strong definition of SSP MDPs. First, there exists at least one proper policy, i.e., one that will eventually terminate in a goal state irrespectively of the starting state (see Definition 2.16). Second, all improper policies incur an infinite cost for those states from which the agent does not reach the goal.

There are two fundamental approaches to solve MDPs optimally. The first is based on iterative techniques that use dynamic programming, whereas the other formulates an MDP as a linear program. Iterative dynamic programming approaches are relatively more popular, since they are amenable to a wider variety of optimizations. Optimal dynamic programming algorithms scale better, hence this chapter focuses primarily on those. We will briefly visit LP solutions in Section 3.7.

We start by describing a simple brute-force algorithm in the next section. An important sub-component of this algorithm is Policy Evaluation, which we discuss in Section 3.2. The most popular and fundamental algorithms for solving MDPs are Policy Iteration (Section 3.3) and Value Iteration (Section 3.4). Two important approaches to optimize these fundamental algorithms are prioritization and partitioning, which are described in Sections 3.5 and 3.6. We end the chapter with important subclasses of MDPs and their properties.

3.1   A BRUTE-FORCE ALGORITHM

Let us begin the study of solving an MDP by considering a simple enumeration algorithm. This algorithm will also expose the complexities in designing an MDP solver. We first observe that there is a finite space of solutions: since there are a total of $|\mathcal{A}|$ actions that can be returned for each state, there are a total of $|\mathcal{S}|^{|\mathcal{A}|}$ possible policies. This leads us to a simple algorithm: enumerate and evaluate all policies π in a brute-force manner and return the best policy π^*.

To implement this brute-force algorithm we need to formalize the notion of evaluating a policy and design algorithms for it. The evaluation of a policy π is equivalent to computing the value function $V^\pi(s)$ for each state s (analogous to Definition 2.7):

Definition 3.2   Value Function of a Policy for an SSP MDP. The *value of a state s under a policy π* is the total expected cost that an agent, starting out in s, incurs before it reaches a goal $g \in \mathcal{G}$.

Observe that $V^\pi(s)$ can even be infinite, if the agent starting in s never reaches a goal. For now we assume that π is a proper policy. We revisit this assumption at the end of Section 3.3.

Analogous to Definition 2.8, an optimal policy π^* for an SSP MDP is one that has minimum value (V^*) for each state, i.e., $\forall s \in \mathcal{S}, \forall \pi, V^*(s) \leq V^\pi(s)$. We now direct our attention to algorithms for evaluating a given policy, i.e., computing V^π.

3.2   POLICY EVALUATION

We first consider special cases for policy evaluation. If all actions in a given policy are deterministic (no probabilistic outcomes), then there is exactly one *path* to a goal from each state. $V^\pi(s)$ is simply the sum of costs of all actions on the path starting from s.

We can visualize a policy through its policy graph, in which there is exactly one edge out of each state. This edge represents the action recommended by the policy for this state. The edge leads to an outcome node (denoted as square boxes in Figure 3.1). This has outgoing edges to different states, one for each non-zero probability transition. For a deterministic MDP, each policy graph has exactly one path from each state to a goal and is easy to evaluate.

By introducing probabilistic actions in a policy, two sources of complexity arise. First, because the action outcomes are not in our control, instead of deterministic path costs we need to compute

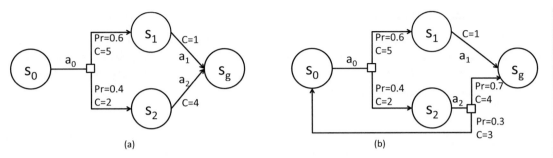

Figure 3.1: Example policy graphs. (a) Acyclic, (b) Cyclic. Evaluating cyclic policy graphs requires solving a system of linear equations or iterative algorithms.

expected path costs. Even here, if the policy graph is *acyclic* (no state can be repeated while executing the policy) the computation is straightforward – instead of the total cost we compute the total expected cost starting from goals to all other states in reverse topological order.

Example: Figure 3.1(a) illustrates an acyclic policy graph for policy π: $\pi(s_0) = a_0, \pi(s_1) = a_1, \pi(s_2) = a_2$. The only probabilistic action a_0 has two outcomes s_1 and s_2 with probabilities 0.6 and 0.4 respectively. The costs are $C(s_0, a_0, s_1) = 5, C(s_0, a_0, s_2) = 2, C(s_1, a_1, g) = 1, C(s_2, a_2, g) = 4$. The only goal state in the domain is g.

 To compute V^π we notice that states s_1 and s_2 reach the goal in one step, and therefore, their values are based on the costs of actions a_1 and a_2: $V^\pi(s_1) = 1, V^\pi(s_2) = 4$. Based on these, we can compute $V^\pi(s_0)$: with probability 0.6 the agent will reach s_1, incur an immediate cost of 5 and a long-term cost of 1 more – so a total cost of 6; similarly with probability 0.4 the total cost will be 2+4 = 6. Thus, the expected cost $V^\pi(s_0) = 0.6 \times 6 + 0.4 \times 6 = 6$. □

 A single value propagation in reverse topological order works for acyclic policy graphs; however, MDPs in general can have *cyclic* policy graphs. This is the second and most important source of complexity for evaluating a policy in an MDP.

Example: Figure 3.1(b) illustrates a cyclic policy graph. Most actions are similar to the previous example, except that a_2 is also probabilistic. With probability 0.7 it reaches the goal and with probability 0.3 it goes back to state s_0.

 Revisiting a state while executing a policy is not uncommon in real domains. Imagine a robot trying to grab the handle of a cup – with some probability it may succeed, or else it may fail, landing back into the original state.

 It is easy to see that in this graph a single value propagation sweep does not work, since the value of s_2 depends on s_0 (as that is a successor) and the value of s_0 depends on s_2 (since it is a successor of a successor). This example illustrates that evaluating a general MDP policy is a subtle matter. □

3.2.1   POLICY EVALUATION BY SOLVING A SYSTEM OF EQUATIONS

To evaluate a cyclic policy, we write a system of linear equations for computing V^π. The basic intuition remains the same: the long-term expected cost in a state depends on two components, the immediate cost incurred in this state and the long-term expected cost incurred starting from the successor states. Mathematically, this translates to:

$$
\begin{aligned}
V^\pi(s) &= 0 && \text{(if } s \in \mathcal{G}) \\
&= \sum_{s' \in \mathcal{S}} \mathcal{T}(s, \pi(s), s') \big[\mathcal{C}(s, \pi(s), s') + V^\pi(s') \big] && \text{(otherwise)}
\end{aligned}
\tag{3.1}
$$

The solution to these equations directly computes the value function of π, since it has $V^\pi(s)$ as the variables. There are a total of $|\mathcal{S}|$ variables, one for each state, so the system can be solved in $O(|\mathcal{S}|^3)$ time using Gaussian Elimination or other algorithms.

Example: Following up on our example of Figure 3.1(b), we can write the following system of equations:

$$
\begin{aligned}
V^\pi(g) &= 0 \\
V^\pi(s_0) &= 0.6(5 + V^\pi(s_1)) + 0.4(2 + V^\pi(s_2)) \\
V^\pi(s_1) &= 1 + V^\pi(g) \\
V^\pi(s_2) &= 0.7(4 + V^\pi(g)) + 0.3(3 + V^\pi(s_0))
\end{aligned}
$$

This set of equations evaluates to $V^\pi(g) = 0$, $V^\pi(s_0) = 6.682$, $V^\pi(s_1) = 1$, and $V^\pi(s_2) = 5.705$. $\square$

3.2.2   AN ITERATIVE APPROACH TO POLICY EVALUATION

We now discuss a second approach to policy evaluation, which is based on iterative value computation algorithms. It is more amenable to approximations and optimizations, as we shall find out later. The key idea is to solve the system of Equations 3.1 using an iterative dynamic programming algorithm.

Our algorithm starts by arbitrarily initializing V^π values for all states – let us call it V_0^π. We then iteratively compute a successive refinement V_n^π using the existing approximation V_{n-1}^π for the successor values. Thus, the algorithm proceeds in iterations and the nth iteration applies the following refinement to the values of all the non-goal states:

$$
V_n^\pi(s) \leftarrow \sum_{s' \in \mathcal{S}} \mathcal{T}(s, \pi(s), s') \big[\mathcal{C}(s, \pi(s), s') + V_{n-1}^\pi(s') \big]
\tag{3.2}
$$

Example: For our running example, let us initialize all V_0^π values to 0. For the first iteration, $V_1^\pi(s_1) = 1 + V_0^\pi(g) = 1 + 0 = 1$, $V_1^\pi(s_0) = 0.6(5 + 0) + 0.4(2 + 0) = 3.8$, and $V_1^\pi(s_2) = 0.7(4 + 0) + 0.3(3 + 0) = 3.7$. Having computed V_1^π, we repeat the same process and compute V_2^π, e.g., $V_2^\pi(s_0) = 5.88$, and so on. $\square$

Notice that this is a dynamic programming procedure that uses the previous layer of values V_{n-1}^{π} to compute the current layer V_n^{π}. In any iterative procedure such as this we need to investigate four crucial properties — convergence, termination, optimality, and running time. We discuss all of these below:

Convergence and Optimality: The first key question is whether the successive values in the iterative algorithm converge to a fixed point or not. We can prove that for any proper policy π, the algorithm converges asymptotically. Not only does it converge to a fixed point, but there is a unique fixed point for the algorithm, and it is the same as the solution to the system of linear equations from Section 3.2.1.[1]

Theorem 3.3   For a Stochastic Shortest Path MDP and a proper policy π, $\forall s \in \mathcal{S}$, $\lim_{n \to \infty} V_n^{\pi}(s) = V^{\pi}(s)$, irrespective of the initialization V_0^{π}.

Termination and Error Bounds: In our fixed-point computation, as n increases, V_n^{π} gets closer and closer to the true value V^{π}. However, it may take infinite iterations to fully converge. For an algorithm to work in practice, we need a termination condition. In response, we define a convergence criterion called ϵ-*consistency*. First, we introduce an auxiliary notion of *residual*.

Definition 3.4   Residual (Policy Evaluation). The *residual at a state s at iteration n* in the iterative policy evaluation algorithm, denoted as residual$_n(s)$, is the magnitude of the change in the value of state s at iteration n in the algorithm, i.e., residual$_n(s) = |V_n^{\pi}(s) - V_{n-1}^{\pi}(s)|$. *The residual at iteration n* is the maximum residual across all states at iteration n of the algorithm, i.e., residual$_n = \max_{s \in \mathcal{S}}$ residual$_n(s)$.

Definition 3.5 ϵ-consistency (Policy Evaluation). The value function V_n^{π} computed at iteration n in iterative policy evaluation is called ϵ-*consistent* if the residual at iteration $n + 1$ is less than ϵ. A state s is called ϵ-*consistent at iteration n* if V_n^{π} is ϵ-consistent at s.

Intuitively, the residual denotes the maximum change in the values of states. Our policy evaluation algorithm terminates when the value function V_n^{π} is ϵ-consistent, i.e., the change in values becomes less than a user-defined ϵ. The pseudo-code for the algorithm is described in Algorithm 3.1.

It is important to note that an ϵ-consistent V_n^{π} may not be ϵ-optimal (i.e., be within ϵ of the fixed-point value V^{π}).

Theorem 3.6   Let Iterative Policy Evaluation be run to ϵ-consistency, i.e., it returns V_n^{π} such that the max residual is less than ϵ. Let $N^{\pi}(s)$ denote the expected number of steps to reach a goal $g \in \mathcal{G}$ from s by following the policy π. Then, V_n^{π} satisfies the following inequality: $\forall s \in \mathcal{S}$: $|V_n^{\pi}(s) - V^{\pi}(s)| < \epsilon N^{\pi}(s)$.

---

[1] The proof depends on the policy evaluation operator being an m-stage contraction and is beyond the scope of this book.

Unfortunately, N^π is difficult to compute for general SSP MDPs, and hence the result is difficult to apply. Known approaches to even bound these require solving equivalently sized MDPs [24]. The special cases of discounted reward problems and SSP MDPs with positive action costs yield much nicer bounds [103; 197]. However, in practice tiny values of ϵ typically do translate to ϵ-consistent value functions that are very close to the fixed point.

Running Time: A single value update for a state requires accessing the values of all the neighbors. Since there can be at most $|\mathcal{S}|$ neighbors, this is an $O(|\mathcal{S}|)$ operation. One full iteration requires this update for all states, making an iteration run in $O(|\mathcal{S}|^2)$ time. Unfortunately, for general SSP MDPs, we cannot provide good bounds on the number of iterations required for ϵ-consistency. Bounds are available for special cases, e.g., when all costs are positive [29].

Algorithm 3.1: Iterative Policy Evaluation

1 //*Assumption: π is proper*
2 initialize V_0^π arbitrarily for each state
3 $n \leftarrow 0$
4 **repeat**
5 $\quad$ $n \leftarrow n + 1$
6 $\quad$ **foreach** $s \in \mathcal{S}$ **do**
7 $\quad\quad$ compute $V_n^\pi(s) \leftarrow \sum_{s' \in \mathcal{S}} \mathcal{T}(s, \pi(s), s') \left[\mathcal{C}(s, \pi(s), s') + V_{n-1}^\pi(s') \right]$
8 $\quad\quad$ compute $\text{residual}_n(s) \leftarrow |V_n^\pi(s) - V_{n-1}^\pi(s)|$
9 $\quad$ **end**
10 **until** $\max_{s \in \mathcal{S}} \text{residual}_n(s) < \epsilon$;
11 return V_n^π

3.3   POLICY ITERATION

Having learned how to evaluate a proper policy, we are in a position to implement our basic enumerate-and-evaluate algorithm in case all policies are proper: enumerate all policies π, evaluate them, and return the best one. There are an exponential number of policies, so the brute-force algorithm is computationally intractable. Policy iteration (PI) [114] makes this basic idea more practical. It replaces the brute-force search by a more intelligent search, so that many policies may not be explored. PI essentially finds an intelligent order for exploring the space of policies, so that our current policy is guaranteed to be better than all the previous policies considered. As a result, PI typically avoids enumerating all policies.

Algorithm 3.2 describes the pseudo-code for PI. At a high level, like the brute-force algorithm, PI starts with evaluating an initial policy π_0. This is accomplished by solving a system of linear equations. Next, based on the value of the current policy it constructs a better one in a policy improvement step. The algorithm keeps alternating between policy evaluation and policy improvement until it cannot improve the policy anymore.

Before we discuss the policy improvement step, we define a few important concepts.

Definition 3.7   Q-value under a Value Function. The *Q-value of state s and action a under a value function V*, denoted as $Q^V(s, a)$, is the one-step lookahead computation of the value of taking a in s under the belief that V is the true expected cost to reach a goal. I.e., $Q^V(s, a) = \sum_{s' \in \mathcal{S}} \mathcal{T}(s, a, s') [\mathcal{C}(s, a, s') + V(s')]$.

Definition 3.8   Action Greedy w.r.t. a Value Function. An action a is *greedy* w.r.t. a value function V in a state s if a has the lowest Q-value under V in s among all actions, i.e., $a = \text{argmin}_{a' \in \mathcal{A}} Q^V(s, a')$.

Any value function V represents a policy π^V that uses only actions greedy w.r.t. V:

Definition 3.9   Greedy Policy. A *greedy policy* π^V for a value function V is a policy that in every state uses an action greedy w.r.t V, i.e., $\pi^V(s) = \text{argmin}_{a \in \mathcal{A}} Q^V(s, a)$.

The policy improvement step computes a greedy policy under $V^{\pi_{n-1}}$. I.e., it first computes the Q-value of each action under $V^{\pi_{n-1}}$ in a given state s. It then assigns a greedy action in s as π_n. The ties are broken arbitrarily, except if $\pi_{n-1}(s)$ still has the best value, in which case that is preferred (lines 7-14 in Algorithm 3.2).

Algorithm 3.2: Policy Iteration

1 initialize π_0 to be an arbitrary proper policy
2 $n \leftarrow 0$
3 **repeat**
4 $\quad$ $n \leftarrow n + 1$
5 $\quad$ Policy Evaluation: compute $V^{\pi_{n-1}}$
6 $\quad$ Policy Improvement:
7 $\quad$ **foreach** *state* $s \in \mathcal{S}$ **do**
8 $\quad\quad$ $\pi_n(s) \leftarrow \pi_{n-1}(s)$
9 $\quad\quad$ $\forall a \in \mathcal{A}$ compute $Q^{(V^{\pi_{n-1}})}(s, a)$
10 $\quad\quad$ $V_n(s) \leftarrow \min_{a \in \mathcal{A}} Q^{(V^{\pi_{n-1}})}(s, a)$
11 $\quad\quad$ **if** $Q^{(V^{\pi_{n-1}})}(s, \pi_{n-1}(s)) > V_n(s)$ **then**
12 $\quad\quad\quad$ $\pi_n(s) \leftarrow \text{argmin}_{a \in \mathcal{A}} Q^{(V^{\pi_{n-1}})}(s, a)$
13 $\quad\quad$ **end**
14 $\quad$ **end**
15 **until** $\pi_n == \pi_{n-1}$;
16 return π_n

PI is a much more efficient version of our basic enumerate-and-evaluate algorithm, since we can prove that each policy improvement step is guaranteed to improve the policy as long as the original π_0 was proper [21]. Thus, π_n monotonically improves, and that is sufficient to guarantee that PI converges, yielding a policy that cannot be improved further. Moreover, it converges to an optimal policy in a finite number of iterations, since there are a finite number of distinct policies.

Theorem 3.10   Policy Iteration for an SSP MDP (initialized with a proper policy π_0) successively improves the policy in each iteration, i.e., $\forall s \in \mathcal{S}$, $V^{\pi_n}(s) \leq V^{\pi_{n-1}}(s)$, and converges to an optimal policy π^* [21].

3.3.1   MODIFIED POLICY ITERATION

A related algorithm is *Modified Policy Iteration* [198; 237]. This algorithm is based on the same principle as policy iteration, except that it uses the iterative procedure of Algorithm 3.1 for policy evaluation. Moreover, instead of arbitrary value initialization $V_0^{\pi_n}$ for policy π_n (line 1 of Algorithm 3.1), Modified PI uses the final value function from the previous iteration $V^{\pi_{n-1}}$.

One version of this algorithm runs the iterative procedure till ϵ-consistency each time (as described in Section 3.2.2). Of course, this is potentially wasteful, since the overall algorithm does not depend upon evaluating the policy exactly. Rather, the value function needs to be improved just enough so that the next better policy can be obtained in the policy improvement step.

Alternative versions of Modified PI run iterative policy evaluation for a pre-defined number of iterations. This speeds up the implementation considerably. In fact, convergence results still apply under additional restrictions [21].

Theorem 3.11   Modified Policy Iteration, initialized with a proper policy π_0, converges to an optimal policy π^*, as long as each policy improvement step performs at least one iteration of iterative policy evaluation, and the initialization $V_0^{\pi_0}$, in iterative policy evaluation, satisfies the condition: $\forall s \in \mathcal{S}$, $\min_{a \in \mathcal{A}} \sum_{s' \in \mathcal{S}} \mathcal{T}(s, a, s')[\mathcal{C}(s, a, s') + V_0^{\pi_0}(s')] \leq V_0^{\pi_0}(s)$.

3.3.2   LIMITATIONS OF POLICY ITERATION

The convergence of PI (and Modified PI) hinges on the first policy being proper. This is because if the first policy π_0 is not proper, the policy evaluation step would diverge. While it may be possible in specific cases to find a seed proper policy using other means, this is not so for the general case. This is a significant drawback of the algorithm. We now discuss another fundamental algorithm, value iteration, which does not suffer from this drawback.

3.4   VALUE ITERATION

Value Iteration (VI) forms the basis of most of the advanced MDP algorithms that we discuss in the rest of the book. It was originally proposed by Richard Bellman and dates back to 1957 [18].

Value iteration takes a complementary perspective to policy iteration. Policy iteration can be visualized as searching in the policy space and computing the current value of a state based on the current policy. Value iteration switches the relative importance of policies and values. It searches directly in the value function space, and computes the current policy based on the current values.

3.4.1 BELLMAN EQUATIONS

VI is based on the set of Bellman equations (Equations 3.3), which mathematically express the optimal solution of an MDP. They provide a recursive expansion to compute the optimal value function V^*, which gives the minimum expected cost to reach a goal starting from a state. Closely related (and a special case of Definition 3.7) is another useful concept, the optimal Q-value of a state-action pair.

Definition 3.12 Optimal Q-value of a state-action pair. The *optimal Q-value of state s and action a*, denoted as $Q^*(s, a)$, is defined as the minimum expected cost to reach a goal starting in state s if the agent's first action is a.

Equivalently, $Q^*(s, a)$ is the expected cost to first execute a in state s, and then follow an optimal policy thereafter.

$$
\begin{aligned}
V^*(s) &= 0 && (\text{if } s \in \mathcal{G}) \\
&= \min_{a \in \mathcal{A}} Q^*(s, a) && (s \notin \mathcal{G}) \\
Q^*(s, a) &= \sum_{s' \in \mathcal{S}} \mathcal{T}(s, a, s') \left[\mathcal{C}(s, a, s') + V^*(s') \right] && (3.3)
\end{aligned}
$$

Notice that this equation is merely a restatement of the Optimality Principle for SSP MDPs (Equation 2.8). The equations can be easily understood in terms of the decision-making of an optimal agent. If the agent is already at a goal, it does not need to execute any action, and the optimal value is zero. For all other states, the agent needs to pick the best action, i.e., an action that minimizes the long-term expected cost to reach a goal. The definition of $Q^*(s, a)$ computes the value for the case that the agent takes a single suboptimal step – the first action; after that, it executes an optimal policy. Thus, $\operatorname{argmin}_{a \in \mathcal{A}} Q^*(s, a)$ represents the agent finding an immediate optimal action and the min-value represents the best value achievable in state s, which is $V^*(s)$. Another direct corollary of the Optimality Principle is the following theorem.

Theorem 3.13 The Bellman equations have a unique solution, which is also the value of all optimal policies π^* for the SSP MDP.

Contrast the Bellman equations with Equation 3.1. When we were given a policy π and our objective was just to evaluate it, it was a simple system of linear equations. However, Bellman equations represent computing the optimal value (implicitly equivalent to finding an optimal policy and computing its value) and, thus, have an additional minimization over all actions. This min makes the equations *non-linear*, and much harder to solve.

Before we discuss the algorithm to solve these equations, it is important to note that any policy greedy w.r.t V^* is optimal. Given V^*, we can easily compute π^* by choosing, for every state, an action greedy w.r.t. V^* in that state (see Definition 3.9). Hence, it suffices to solve the set of Bellman equations to solve an MDP.

3.4.2   THE VALUE ITERATION ALGORITHM

VI computes a solution to the system of Bellman equations via successive refinements, using ideas similar to the iterative policy evaluation of Section 3.2.2. The key idea is to successively approximate V^* with a V_n function, such that the sequence of V_ns converges to V^* in the limit as n tends to infinity.

The algorithm proceeds in iterations (Algorithm 3.3). It first initializes all values V_0 arbitrarily. In the nth iteration, it makes a *full sweep*, i.e., iterates over all states, computing a new approximation V_n for the state values using the successor values from the previous iteration (V_{n-1}) as:

$$V_n(s) \leftarrow \min_{a \in \mathcal{A}} \sum_{s' \in \mathcal{S}} \mathcal{T}(s, a, s') \left[\mathcal{C}(s, a, s') + V_{n-1}(s') \right]. \tag{3.4}$$

Compare this refinement step with Equation 3.2 used for policy evaluation. Since policy evaluation evaluates a given policy, the action in each state is known and therefore, there is no min component in that equation. In contrast, since no *a priori* policy is known in VI, it has the additional non-linearity of the min over all actions. This particular computation of calculating a new value based on successor-values is known as a *Bellman backup* or a *Bellman update*.

Definition 3.14   **Bellman backup.** A *Bellman backup at state s w.r.t. a value function V* computes a new value at s by backing up the successor values $V(s')$ using Equation 3.4.

Example: We illustrate the execution of VI with the example in Figure 3.2. There are six states in this domain. One to two actions are applicable in each state. All costs are 1 except for actions a_{40} and a_{41}, which cost 5 and 2 respectively. All actions are deterministic except for a_{41}. Say, VI initializes all V_0 values as the distance from the goal, i.e., the values are 3, 3, 2, 2, and 1 for s_0, s_1, s_2, s_3, and s_4 respectively. Also, let the order in which we back up states be $s_0, s_1, \ldots, s_4$.

In iteration 1, VI will sweep over all states and perform Bellman backups. For example, $V_1(s_0) = \min\{1 + V_0(s_1), 1 + V_0(s_2)\} = 3$. Similarly, V_1 for s_1, s_2, and s_3 is 3, 2, and 2. For s_4, $Q_1(s_4, a_{41}) = 2 + 0.6 \times 0 + 0.4 \times 2 = 2.8$ and $Q_1(s_4, a_{40}) = 5$. So, $V_1(s_4) = \min\{2.8, 5\} = 2.8$. This completes the first iteration. These values will be further backed up in the second iteration. Figure 3.3 lists the results of VI on the domain. Notice that as the number of iterations increases, the values tend to stabilize.   □

VI is easily understood as an iterative, fixed-point computation algorithm — Bellman equations specify a set of recursive equations, and VI is a method to approach the fixed-point of these

---

Algorithm 3.3: Value Iteration

1 initialize V_0 arbitrarily for each state
2 $n \leftarrow 0$
3 **repeat**
4      $n \leftarrow n + 1$
5   **foreach** $s \in \mathcal{S}$ **do**
6   compute $V_n(s)$ using Bellman backup at s
7   compute $\text{residual}_n(s) = |V_n(s) - V_{n-1}(s)|$
8   **end**
9 **until** $\max_{s \in \mathcal{S}} \text{residual}_n(s) < \epsilon$;
10 return greedy policy: $\pi^{V_n}(s) = \text{argmin}_{a \in \mathcal{A}} \sum_{s' \in \mathcal{S}} \mathcal{T}(s, a, s') [\mathcal{C}(s, a, s') + V_n(s')]$

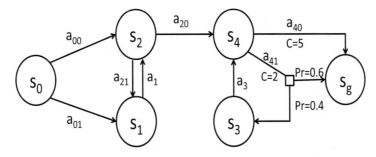

Figure 3.2: Example MDP domain. All unstated costs are 1.

equations using successive approximation. VI is also a dynamic programming algorithm — the whole layer of values V_n is stored for the next iteration.

VI can also be thought of as a message passing algorithm to achieve a global information flow. It passes only local messages (sent only between neighbors through the Bellman backup) and computes a globally optimal value function.

Finally, VI also relates to the shortest path algorithms in graph theory. Regular graphs consist of deterministic edges. General MDPs, on the other hand, can be represented as *AND/OR graphs*, where a state is an OR node and an edge leads to an AND node (shown as square boxes in the domain of Figure 3.2). AND/OR graphs are so named, because the OR nodes may be solved by any of a number of alternative ways, but an AND node can be solved only when all of immediate subproblems are solved. In MDPs, the outcome nodes are AND nodes, since once an action is selected we need to evaluate all possible successors. A state is an OR node, since the agent needs to select exactly one outgoing edge (action). VI is a method to compute the shortest path in AND/OR graphs. It can be seen as the stochastic extension of the Bellman-Ford algorithm for shortest path computation in graphs. If all outcomes have probability 1, indeed, VI reduces to the Bellman-Ford algorithm [55].

n	$V_n(s_0)$	$V_n(s_1)$	$V_n(s_2)$	$V_n(s_3)$	$V_n(s_4)$
0	3	3	2	2	1
1	3	3	2	2	2.8
2	3	3	3.8	3.8	2.8
3	4	4.8	3.8	3.8	3.52
4	4.8	4.8	4.52	4.52	3.52
5	5.52	5.52	4.52	4.52	3.808
10	5.9232	5.9232	4.96928	4.96928	3.96928
20	5.99921	5.99921	4.99969	4.99969	3.99969

Figure 3.3: Value Iteration run on the example of Figure 3.2.

3.4.3   THEORETICAL PROPERTIES

Recall that any iterative algorithm can be characterized by four key properties – convergence, termination, optimality and running time.

Convergence and Optimality: VI values converge to the optimal in the limit. Unlike PI, which requires an initial proper policy, VI converges without restrictions.

Theorem 3.15   For a Stochastic Shortest Path MDP, $\forall s \in \mathcal{S}, \lim_{n \to \infty} V_n(s) = V^*(s)$, irrespective of the initialization V_0.

Termination and Error Bounds: The termination condition for VI mimics that of the iterative policy evaluation algorithm. We can adapt the notion of the residual of a state from Definition 3.4 (also called *Bellman error* for VI), which denotes the magnitude of change in the values between two successive iterations. Below we define slightly more general notions of a residual and ϵ-consistency, which we will use in the rest of the book.

Definition 3.16   Residual (Bellman Backup). In an SSP MDP, *the residual at a state s w.r.t. a value function V*, denoted as $Res^V(s)$, is the magnitude of the change in the value of state s if Bellman backup is applied to V at s once, i.e., $Res^V(s) = |V(s) - \min_{a \in A}(\sum_{s' \in \mathcal{S}} \mathcal{T}(s, a, s')[C(s, a, s') + V(s')])|$. *The residual w.r.t. a value function V*, denoted as Res^V, is the maximum residual w.r.t. V across all states if Bellman backup is applied to V at each state once, i.e., $Res^V = \max_{s \in \mathcal{S}} Res^V(s)$.

Definition 3.17 ϵ-consistency (Bellman Backup). A state s is called *ϵ-consistent w.r.t. a value function V* if the residual at s w.r.t. V is less than ϵ, i.e., $Res^V(s) < \epsilon$. A value function V is called *ϵ-consistent* if it is ϵ-consistent at all states, i.e., if $Res^V < \epsilon$.

We terminate VI if all residuals are small. Note that once all residuals are less than ϵ, in case we performed more iterations, they will remain less than ϵ for all subsequent iterations. As usual, if all the residuals are small, we are typically quite close to the optimal. The following bounds apply for the converged value functions [21].

Theorem 3.18   Let Value Iteration be run to ϵ-consistency, i.e., it returns V_n such that the max residual is less than ϵ. Let $N^*(s)$ and $N^{\pi_n}(s)$ denote the expected number of steps to reach a goal $g \in \mathcal{G}$ from s by following the optimal policy and the current greedy policy respectively. Then, V_n satisfies the following inequality: $\forall s \in \mathcal{S} : |V_n(s) - V^*(s)| < \epsilon \cdot \max\{N^*(s), N^{\pi^{V_n}}(s)\}$.

The same bound also applies to the value of a greedy policy π^{V_n}. Unfortunately, N^* and N^π are difficult to compute for general SSP MDPs, and hence the result is difficult to apply. Known approaches to even bound these require solving equivalently sized MDPs [24]. The special cases of discounted reward problems yield much nicer bounds (see Section 3.8). Better bounds are also available for the case of positive action costs [103]. In practice, VI typically obtains optimal or near-optimal policies for small ϵ values, and the practitioners are happy with ϵ-consistent value functions as the solution to their problems.

Running Time: A single Bellman backup runs in worst-case $O(|\mathcal{S}||\mathcal{A}|)$ time, since it requires iterating over all actions and all successor states. A single iteration requires $|\mathcal{S}|$ backups, so a complete iteration requires $O(|\mathcal{S}|^2|\mathcal{A}|)$. We can bound the number of iterations for special cases of SSP MDPs (e.g., when the all costs are positive and the initial value function is a lower bound on the optimal [29]), but non-trivial bounds do not exist for general SSP MDPs.

Monotonicity: VI satisfies another useful property, called *monotonicity*.

Definition 3.19   **Monotonicity.** An operator $T : (\mathcal{S} \to [-\infty, \infty]) \to (\mathcal{S} \to [-\infty, \infty])$, which applies on a value function to obtain a new value function, is *monotonic* if $\forall V_1, V_2, \ V_1 \leq V_2 \Rightarrow T(V_1) \leq T(V_2)$.

In other words, if a value function V_1 is componentwise greater (or less) than another value function V_1 then the same inequality holds true between $T(V_1)$ and $T(V_2)$, i.e., the value functions that are obtained by applying this operator T on V_1 and V_2. We can prove that the Bellman backup operator used in VI is monotonic [103]. As a corollary, if a value function (V_k) is a lower (or upper) bound on V^* for all states, then all intermediate value functions thereafter $(V_n, n > k)$ continue to remain lower (respectively, upper) bounds. This is because V^* is the fixed point of the Bellman backup operator.

Theorem 3.20   In Value Iteration, if $\forall s \in \mathcal{S}, V_k(s) \leq V^*(s)$, then $\forall s \in \mathcal{S}, n > k, \ V_n(s) \leq V^*(s)$. Similarly, if $\forall s \in \mathcal{S}, V_k(s) \geq V^*(s)$, then $\forall s \in \mathcal{S}, n > k, \ V_n(s) \geq V^*(s)$.

3.4.4 ASYNCHRONOUS VALUE ITERATION

One of the biggest drawbacks of VI (also called *synchronous VI*) is that it requires full sweeps of the state space. The state spaces are usually large for real problems, making VI impractical. However, not all states may be equally important. Moreover, Bellman backups at some states may be more useful, because they may change the values considerably, and less for other states, where the values may have more or less converged. These intuitions go a long way in optimizing the value function computation. To enable such algorithms, asynchronous VI relaxes the requirement that all states need to be backed up in each iteration.

Algorithm 3.4: Asynchronous Value Iteration

1 initialize V arbitrarily for each state
2 **while** $Res^V > \epsilon$ **do**
3 $\quad$ select a state s
4 $\quad$ compute $V(s)$ using a Bellman backup at s
5 $\quad$ update $Res^V(s)$
6 **end**
7 return greedy policy π^V

Algorithm 3.4 provides pseudocode for the most generic version of *Asynchronous VI*, where in the inner loop a state is selected and a Bellman backup applied over it. Convergence of Asynchronous VI (in the limit) requires an additional restriction so that no state gets starved, i.e., all states are backed up an infinite number of times (infinitely often). With this restriction, for Asynchronous VI, we can prove that $\forall s \in \mathcal{S}, \lim_{\epsilon \to 0} V(s) = V^*(s)$ [23].

As usual, in practice we cannot run the algorithm forever (ϵ=0), so a termination condition (line 2), similar to that of Value Iteration, is employed. It checks whether the current value function is ϵ-consistent for some small nonzero ϵ.

A special case of Asynchronous VI is the *Gauss-Seidel version of VI*, where the states are still selected in a round-robin fashion [23]. Notice that even though this makes sweeps over states in a fixed order, it is still different from VI. In VI, the successor values that are being backed up are from the previous iteration. However, in the Gauss-Seidel version, all interim computation of the current iteration is available to a Bellman backup. That is, the successor values could be from this iteration itself.

The general version of Asynchronous VI forms the basis of all optimizations to VI, since it allows the flexibility to choose a backup order intelligently — the main focus of the rest of the chapter.

3.5 PRIORITIZATION IN VALUE ITERATION

Consider the VI example from Figure 3.2. Notice that in the first few iterations, the value of state s_0 is not changing. Nonetheless, VI still wastes its time performing Bellman backups for s_0 in these iterations. This section investigates reducing such wasteful backups by choosing an intelligent backup order. The key idea is to additionally define a *priority* of each state s, representing an estimate of how helpful it is to back up s. Higher-priority states are backed up earlier.

Algorithm 3.5: Prioritized Value Iteration

1 initialize V
2 initialize priority queue q
3 **repeat**
4   select state $s' = q.pop()$
5   compute $V(s')$ using a Bellman backup at s'
6   **foreach** *predecessor s of s', i.e., $\{s|\exists a[\mathcal{T}(s, a, s') > 0]\}$* **do**
7   compute priority(s)
8          $q.push(s, \text{priority}(s))$
9   **end**
10 **until** *termination*;
11 return greedy policy π^V

When can we avoid backing up a state s? An obvious answer is: when none of the successors of s have had a change in value since the last backup — this would mean that backing up s wil not result in a change of its value. In other words, a change in the value of a successor is an indicator for changing the priority of a state.

We formalize this intuition in Algorithm 3.5 with a general Prioritized Value Iteration scheme (adapted from [246]). It implements an additional priority queue that maintains the priority for backing up each state. In each step, a highest-priority state is popped from the queue and a backup is performed on it (lines 4-5). The backup results in a potential change of priorities for all the predecessors of this state. The priority queue is updated to take into account the new priorities (lines 6-9). The process is repeated until termination (i.e., until $Res^V < \epsilon$).

Convergence: Prioritized VI is a special case of Asynchronous VI, so it converges under the same conditions — if all states are updated infinitely often. However, because of the prioritization scheme, we cannot guarantee that all states will get updated an infinite number of times; some states may get *starved* due to their low priorities. Thus, in general, we cannot prove the convergence of Prioritized VI. In fact, Li and Littman [151] provide examples of domains and priority metrics where Prioritized VI does not converge. We can, however, guarantee convergence to the optimal value function, if we simply interleave synchronous VI iterations within Prioritized VI. While this is an additional overhead, it mitigates the lack of theoretical guarantees. Moreover, we can prove convergence without this interleaving for specific priority metrics (such as prioritized sweeping).

Researchers have studied different versions of this algorithm by employing different priority metrics. We discuss these below.

3.5.1   PRIORITIZED SWEEPING

Probably the first and most popular algorithm to employ an intelligent backup order is Prioritized Sweeping (PS) [183]. PS estimates the expected change in the value of a state if a backup were to be performed on it now, and treats this as the priority of a state. Recall that $Res^V(s')$ denotes the change in the value of s' after its Bellman backup. Let s' was backed up with a change in value $Res^V(s')$ then PS changes the priority of all predecessors of s'. If the highest-probability transition between s and s' is p then the maximum residual expected in the backup of s is $p \times Res^V(s')$. This forms the priority of s after the backup of s':

$$\text{priority}_{PS}(s) \leftarrow \max\left\{\text{priority}_{PS}(s), \max_{a\in\mathcal{A}}\left\{\mathcal{T}(s, a, s')Res^V(s')\right\}\right\} \tag{3.5}$$

For the special case, where the state is a successor of itself, the outer max step is ignored: $\text{priority}(s') \leftarrow \max_{a\in\mathcal{A}}\{\mathcal{T}(s, a, s')Res^V(s')\}$.

Example: Continuing with example in Figure 3.2, say, we first back up all states in the order s_0 to s_4 (as in the first iteration of VI). So far, the only state that has changed value is s_4 (residual = 1.8). This increases the priorities of s_2 and s_3 to be 1.8, so one of them will be the next state to be picked up for updates. Note that the priority of s_0 will remain 0, since no change in its value is expected. □

Convergence: Even though PS does not update each state infinitely often, it is still guaranteed to converge under specific conditions [151].

Theorem 3.21   Prioritized Sweeping converges to the optimal value function in the limit, if the initial priority values are non-zero for all states $s \in \mathcal{S}$.

This restriction is intuitive. As an example, if all predecessors of goals have initial priorities zero (s_4 in our example), a goal's base values will not be propagated to them and hence, to no other states. This may result in converging to a suboptimal value function.

The practical implementations of PS for SSP MDPs make two additional optimizations. First, they initialize the priority queue to contain only the goal states. Thus, the states start to get backed up in the reverse order of their distance from goals. This is useful, since values flow from the goals to the other states. Second, for termination, they update the priorities only if $\text{priority}_{PS}(s) > \epsilon$. Thus, the algorithm does not waste time in backing up states where the residual is expected to be less than ϵ.

Generalized Prioritized Sweeping: Generalized PS [4] suggests two priority metrics. The first is a modification of PS where priorities are not maxed with the prior priority, rather, added as follows:

$$\text{priority}_{GPS1}(s) \leftarrow \text{priority}_{GPS1}(s) + \max_{a\in\mathcal{A}}\left\{\mathcal{T}(s, a, s')Res^V(s')\right\} \tag{3.6}$$

This makes the priority of a state an *overestimate* of its residual. The original intent of GPS1's authors was that this would avoid starvation, since states will have a higher priority than their residual. But, as was later shown, this scheme can potentially starve other states [151].

The second metric (GPS2) computes the actual residual of a state as its priority:

$$
\begin{aligned}
\text{priority}_{GPS2}(s) \quad &\leftarrow \quad Res^V(s) \\
&= \quad \left| \sum_{s' \in \mathcal{S}} \mathcal{T}(s, a, s') \left[C(s, a, s') + V(s') \right] - V(s) \right|
\end{aligned}
\tag{3.7}
$$

The implementation of the GPS2 priority scheme differs from our pseudocode for Prioritized VI in one significant way — instead of estimating the residual, it uses its *exact* value as the priority. In other words, one needs to perform the backup to even compute the priority function. The effect is that value propagation happens at the time a state is *pushed* in the queue (as opposed to when it is popped, in the case of PS). This particular metric does not require interleaved synchronous VI iterations for convergence [151].

3.5.2   IMPROVED PRIORITIZED SWEEPING

PS successfully incorporates the intuition that if the value of a successor, $V(s')$, has changed, then the values of the predecessor s may have changed also, so its backup needs to be scheduled. While this insight is accurate and useful, it is also somewhat myopic. At the global level, we wish to iterate these values to their fixed-points. If $V(s')$ is still in flux, then any further change in it will require additional propagations to s.

For example, Figure 3.3 shows that the values of s_3 and s_4 converge slowly due to their interdependencies. And each iteration propagates their intermediate values to $s_0 - s_2$. These are essentially wasted computations. We could save a lot of computation if we first allowed s_3 and s_4 to converge completely and then propagated their values to $s_0 - s_2$ just once. Of course, since we are dealing with cyclic graphs, this may not always be possible. However, we could at least incorporate this insight into our prioritization scheme.

In SSP MDPs, values propagate from goals backwards to other states. To include proximity to a goal as a criterion in prioritization, Improved Prioritized Sweeping[2] [175] defines the following priority metric (only defined for non-goals, $V(s) > 0$):

$$
\text{priority}_{IPS}(s) \leftarrow \frac{Res^V(s)}{V(s)}
\tag{3.8}
$$

The division by $V(s)$ trades off residual and proximity to a goal. Since the states closer to a goal will have smaller $V(s)$, thus, their priority will be higher than states farther away. Slowly, as their residual reduces, states farther away become more important and they get backed up. Note that

---

[2]Their original paper did not perform Bellman backups, and presented a different expansion-based algorithm. Their metric in the context of Prioritized VI was discussed first in [60].

the implementation of this algorithm is similar to that of the second variant of Generalized PS – the states are backed up when they are pushed in the queue rather than when they are popped.

3.5.3 FOCUSED DYNAMIC PROGRAMMING

In the special case when the knowledge of a start state s_0 is available (discussed in detail in Chapter 4), the prioritization scheme can also take into account the relevance of a state with respect to the solution from the start state. *Focused Dynamic Programming* [87; 88] defines the priority metric priority$_{FDP}(s) \leftarrow h(s) + V(s)$. Here, $h(s)$ is a lower bound on the expected cost of reaching s from s_0. Note that for this algorithm, lower priority values are backed up first.

3.5.4 BACKWARD VALUE ITERATION

In all these prioritized algorithms the existence of a priority queue is a given. However, sometimes the overhead of a priority queue can be significant. Especially, for large domains, where states have a bounded number of successors, a Bellman backup takes constant time but each priority queue update may take $O(log|\mathcal{S}|)$ time, which can be very costly. Wingate & Seppi [246] provide an example of a domain where Generalized PS performs half as many backups as VI but is 5-6 times slower, with the bulk of the running time attributed to priority queue operations.

In response, the Backward Value Iteration (BVI) algorithm [60] suggests performing backups simply in the reverse order of the distance from a goal using a FIFO queue (as opposed to a priority queue). Another optimization, instead of backing up all states in this fashion, backs up only the states that are in the greedy subgraph starting from a goal state (backward).

Algorithm 3.6 elaborates on the BVI algorithm. In each iteration, the queue starts with the goal states and incrementally performs backups in the FIFO order. A state is never reinserted into the queue in an iteration (line 11). This guarantees termination of a single iteration. Finally, note line 10. It suggests backing up only the states that are in the current greedy policy. This is in contrast with other PS algorithms that back up a state if they are a predecessor of the current backed up state, irrespective of whether the action is currently greedy or not. It is not clear how important this modification is for the performance of BVI, though similar ideas have been effective in other algorithms, e.g., ILAO* (discussed in Section 4.3.2).

No convergence results are known for BVI, though our simple trick of interleaving VI iterations will guarantee convergence. In practice, BVI saves on the overhead of priority queue implementation, and in many domains it also performs far fewer backups than other PS algorithms. The results show that it is often more efficient than other PS approaches [60].

3.5.5 A COMPARISON OF PRIORITIZATION ALGORITHMS

Which prioritization algorithm to choose? It is difficult to answer this question, since different algorithms have significantly different performance profiles and they all work very well for some domains.

Algorithm 3.6: Backward Value Iteration

1   initialize V arbitrarily
2   **repeat**
3      $visited = \emptyset$
4      $\forall g \in \mathcal{G}, q.push(g)$
5   **while** $q \neq \emptyset$ **do**
6          $s' = q.pop()$
7          $visited.insert(s')$
8   compute $V(s')$ using a Bellman backup at s'
9   store $\pi(s') = \pi^V(s')$ //greedy action at s'
10   **foreach** $\{s|\mathcal{T}(s, \pi(s), s') > 0\}$ **do**
11   **if** s *not in visited* **then**
12                $q.push(s)$
13   **end**
14   **end**
15   **end**
16 **until** *termination*;
17 return greedy policy π^V

Non-prioritized (synchronous) VI provides very good backup orders in cases where the values of states are highly interdependent. For example, for a fully connected subgraph, one cannot possibly do much better than VI. On such domains, prioritized methods add only computational overhead and no value. On the other hand, if a domain has highly sequential dependencies, then VI with random state orders falls severely short.

PS and Generalized PS obtain much better performance for domains with sequential dependencies, such as acyclic graphs or graphs with long loops. They work great for avoiding useless backups but tend to be myopic, since they do not capture the insight that one can save on propagating all intermediate information to farther-away states by first letting the values closer to goals stabilize.

Improved PS suggests one specific way to trade off between proximity to a goal and residual as means of prioritization. However, this particular tradeoff may work really well for some domains, and may give rather suboptimal backup orders for others.

Finally, BVI is a priority-queue-free algorithm, which completely shuns any residual-based prioritization and focuses entirely on ordering based on distance from a goal (in the greedy subgraph). It is shown to perform significantly better than other PS algorithms for several domains, but it may fail for domains with large fan-ins (i.e., with states with many predecessors). An extreme case is SSP MDPs compiled from infinite-horizon discounted-reward problems (refer to the transformation discussed in Section 2.4), where all states have a one-step transition to a goal.

3.6 PARTITIONED VALUE ITERATION

We now focus our attention on another important idea for managing VI computations — the idea of partitioning the state space. We illustrate this again with the example of Figure 3.2. What is the optimal backup order for this MDP? We note that states s_3 and s_4 are mutually dependent, but do not depend on values of $s_0 - s_2$. We will need several iterations for the values of s_3 and s_4 to converge. Once that happens, the values can be backed up to s_1 and s_2, and finally to s_0. This example clearly illustrates that we need to stabilize mutual co-dependencies before attempting to propagate the values further.

We can formalize this intuition with the notion of Partitioned Value Iteration [246] – we create a partitioning of states (in our example, $\{\{s_0\}, \{s_1, s_2\}, \{s_3, s_4\}\}$) makes the best partitioning for information flow) and perform several backups within a partition before focusing attention on states in other partitions.

Additionally, we can also incorporate a prioritization scheme for choosing an intelligent *partition order*. In our example, partitions need to be considered in the order from right to left. All the intuitions from the previous section can be adapted for the partitioned case. However, instead of successor/predecessor states, the notion of successor/predecessor partitions is more relevant here. For partition $\mathfrak{p}'$, we can define another partition $\mathfrak{p}$ to be a *predecessor partition* if some state in $\mathfrak{p}$ is a predecessor of a state in $\mathfrak{p}'$. Formally,

$$\text{PredecessorPartition}(\mathfrak{p}') = \{\mathfrak{p}|\exists s \in \mathfrak{p}, \exists s' \in \mathfrak{p}', \exists a \in \mathcal{A} \ s.t. \ \mathcal{T}(s, a, s') > 0\} \qquad (3.9)$$

Algorithm 3.7 provides pseudocode for a general computation scheme for Partitioned VI. If line 6 only executes one backup per state and the prioritization scheme is round-robin, then Partitioned VI reduces to VI. If all partitions are of size 1 then the algorithm, depending on the priority, reduces to various PS algorithms. Another extreme case is running each partition till convergence before moving to another partition. This version is also known as *Partitioned Efficient Value Iterator (P-EVA)* [243].

Which algorithm should one use within a partition (line 6)? Essentially any algorithm can be used, from VI, PI to any of the prioritized versions. The convergence of Partitioned VI follows the same principles as before. If we can guarantee that the residual of each state is within ϵ, then the algorithm has converged and can be terminated. This can be achieved by interleaving VI sweeps or ensuring any other way that no state/partition starves.

More importantly, how do we construct a partitioning? Different approaches have been suggested in the literature, but only a few have been tried. Intuitively, the partitioning needs to capture the underlying co-dependency between the states – states whose values are mutually dependent need to be in the same partition. If the states have specific structure, it can be exploited to generate partitions. Say, if the states have associated spatial location, then nearby states can be in the same partition. Domain-independent ways to partition states could employ k-way graph partitioning or

---

Algorithm 3.7: Partitioned Value Iteration

1 initialize V arbitrarily
2 construct a partitioning of states $\mathfrak{P} = \{\mathfrak{p}_i\}$
3 (optional) initialize priorities for each $\mathfrak{p}_i$
4 **repeat**
5   select a partition $\mathfrak{p}'$
6   perform (potentially several) backups for all states in $\mathfrak{p}'$
7   (optional) update priorities for all predecessor partitions of $\mathfrak{p}'$
8 **until** *termination*;
9 return greedy policy π^V

---

clustering algorithms [3; 69; 184; 186]. We now discuss a special case of Partitioned VI that employs a specific partitioning approach based on strongly-connected components.

3.6.1 TOPOLOGICAL VALUE ITERATION

Topological Value Iteration (TVI) [59; 66] uses an MDP's graphical structure to compute a partitioning of the state space. It first observes that there is an easy way to construct an optimal backup order in the case of acyclic MDPs [21].

Theorem 3.22 *Optimal Backup Order for Acyclic MDPs.* If an MDP is acyclic, then there exists an optimal backup order. By applying the optimal order, the optimal value function can be found with each state needing only one backup.

The reader may draw parallels with the policy computation example from Figure 3.1(a). In the case of an acyclic graph, we can propagate information from goal states backward and return an optimal solution in a single pass. Of course, the interesting case is that of cyclic MDPs, where such an order does not exist. But we may adapt the same insights and look for an *acyclic partitioning*. If we could find a partitioning where all partitions are linked to each other in an acyclic graph then we could back up partitions in the optimal order and run VI within a partition to convergence.

The key property of this partitioning is that the partitions and links between them form an acyclic graph. That is, it is impossible to go back from a state in a succeeding partition to a state in the preceding partition. Thus, all states within a partition form a *strongly connected component (SCC)*: we can move back and forth between all states in an SCC. However, this property does not hold true across different SCCs.

The computation of SCCs is well studied in the algorithm literature. TVI first converts the MDP graph into a deterministic causal graph, which has a directed edge between two states s and s', if there is a non-zero probability transition between the two: $\exists a \in \mathcal{A}[\mathcal{T}(s, a, s') > 0]$. TVI then applies an algorithm (e.g., the Kosaraju-Sharir algorithm, which runs linear in the size of the graph [55]) to compute a set of SCCs on this graph. These are used as partitions for Partitioned VI. The reverse topological sort on the partitions generates an *optimal* priority order to backup these partitions in the sense that a partition is backed up after all its successors have been run to ϵ-consistency.

Theorem 3.23   Topological Value Iteration for an SSP MDP is guaranteed to terminate with an ϵ-consistent value function, as long as $\epsilon > 0$.

Example: In the example MDP from Figure 3.2, an SCC algorithm will identify the partitions to be $\mathfrak{p}_0 = \{s_0\}$, $\mathfrak{p}_1 = \{s_1, s_2\}$, $\mathfrak{p}_2 = \{s_3, s_4\}$ $\mathfrak{p}_3 = \{g\}$. The reverse topological sort on the partitions will return the backup order as $\mathfrak{p}_3, \mathfrak{p}_2, \mathfrak{p}_1, \mathfrak{p}_0$. Since $\mathfrak{p}_3$ only has the goal state, it does not need backups. TVI will run VI for all states within $\mathfrak{p}_2$ next. Once V_n for s_3 and s_4 converge, they will be propagated to $\mathfrak{p}_1$. The values from $\mathfrak{p}_1$ will propagate to $\mathfrak{p}_0$ to compute $V^*(s_0)$.   □

TVI offers large speedups when a good SCC-based partitioning exists, i.e., each partition is not too large. However, often, such a partitioning may not exist. For example, any reversible domain (each state is reachable from all other states) has only one SCC. For such problems, TVI reduces to VI with an additional overhead of running the SCC algorithm, and is not effective. An extension of TVI, known as *Focused Topological Value Iteration* prunes suboptimal actions and attains good partitioning in a broad class of domains. This algorithm is discussed in Section 4.5.2.

Partitioned VI naturally combines with a few other optimizations. We discuss a few below when describing external-memory algorithms, cache-efficient algorithms, and parallel algorithms.

3.6.2   EXTERNAL MEMORY/CACHE EFFICIENT ALGORITHMS

VI-based algorithms require $O(|\mathcal{S}|)$ memory to store current value functions and policies. In problems where the state space is too large to fit in memory, disk-based algorithms can be developed. However, since disk I/O is a bottleneck, value iteration or general prioritized algorithms may not perform as well. This is because these require random access of state values (or a large number of sweeps over the state space), which are I/O-inefficient operations.

Partitioned VI fits nicely in the external memory paradigm, since it is able to perform extensive value propagation within a partition before focusing attention on the next partition, which may reside on the disk. Thus, Partitioned VI may terminate with much fewer I/O operations.

Partitioned External Memory Value Iteration (PEMVI) [61] partitions the state space so that all information relevant to backing up each partition (transition function of this partition and values of current and all successor partitions) fits into main memory. It searches for such a partitioning via domain analysis.[3] The details of this work are out of the scope of this book. The reader may refer to the original paper for details [62].

An interesting detail concerns PEMVI's backup orders. There are two competing backup orders — one that minimizes the I/O operations and one that maximizes the information flow (reduces number of backups). The experiments show that the two are not much different and have similar running times.

---

[3]The partitioning algorithm investigates the special case when the domains are represented in Probabilistic PDDL [253] representation.

Partitioning is not the only way to implement external-memory MDP algorithms. *External Memory Value Iteration* is an alternative algorithm that uses clever sorting ideas to optimize disk I/Os [82]. In practice, PEMVI with good partitioning performs better than EMVI [61].

Closely associated with the I/O performance of an external-memory algorithm is the cache efficiency of the internal memory algorithm. To our knowledge, there is only one paper that studies cache performance of the VI-based algorithms [244]. It finds that Partitioned VI, in particular, the P-EVA algorithm (with specific priority functions), obtains a much better cache performance. This is not a surprising result, since, by partitioning and running backups till convergence on the current partition, the algorithm naturally accesses the memory with better locality.

We believe that more research on cache efficiency of MDP algorithms is desirable and could lead to substantial payoffs — similar to the literature in the algorithms community (e.g., [147]), one may gain huge speedups due to better cache performance of the algorithms.

3.6.3   PARALLELIZATION OF VALUE ITERATION

VI algorithms are easily parallelizable, since different backups can be performed in parallel. An efficient parallel algorithm is P3VI, which stands for *Partitioned, Prioritized, Parallel Value Iterator*. Breaking the state space into partitions naturally allows for efficient parallelized implementations — different partitions are backed up in parallel.

The actual algorithm design is more challenging than it sounds, since it requires us to construct a strategy to allocate processors to a partition, and also manage the messages between the processors when the value function of a partition has changed, which may result in reprioritizing the partitions. We refer the reader to the original paper [245] for those details.

The stopping condition is straightforward. P3VI terminates when all processors report a residual less than ϵ. Unsurprisingly, the authors report significant speedups compared to naive multi-processor versions of the VI algorithm.

3.7   LINEAR PROGRAMMING FORMULATION

While iterative solution algorithms are the mainstay of this book, no discussion about MDP algorithms is complete without a mention of an alternative solution approach that is based on linear programming. The set of Bellman equations for an SSP MDP can also be solved using the following LP formulation [78] ($\alpha(s) > 0, \forall s$):

$$
\begin{aligned}
\text{Variables} \quad & V^*(s) \quad \forall s \in \mathcal{S} \\
\text{Maximize} \quad & \sum_{s \in \mathcal{S}} \alpha(s) V^*(s) \\
\text{Constraints} \quad & V^*(s) = 0 \quad \text{if } s \in \mathcal{G} \\
& V^*(s) \leq \sum_{s' \in \mathcal{S}} [\mathcal{C}(s, a, s') + \mathcal{T}(s, a, s') V^*(s')] \ \forall s \in \mathcal{S} \setminus \mathcal{G}, a \in \mathcal{A}
\end{aligned}
$$

Here, $\alpha(s)$ are known as the state-relevance weights. For the exact solution these are unimportant and can be set to any positive number (say, to 1). These tend to become more useful for approximation algorithms that we will discuss in later chapters.

The LP formulation is closely related to iterative approaches. For instance, the simplex method to solve this LP with block pivoting is mathematically equivalent to policy iteration. For more details see [197].

In theory, LPs can be solved in polynomial time. However, in practice, LPs computing the exact solution of MDPs are found to be slower than the VI-extensions discussed in this chapter. However, LPs are more amenable to a specific kind of approximation, where the value of a state is approximated via a sum of basis functions. We revisit this in Section 6.4.2.

3.8 INFINITE-HORIZON DISCOUNTED-REWARD MDPS

Infinite-horizon MDPs maximizing discounted rewards are commonly studied in the AI literature (see Definition 2.14). Although they are a special case of SSP MDPs (discussed in Chapter 2), it is worth discussing them separately, since some SSP results simplify for this special case. Moreover, some algorithms for SSP problems are less effective for these MDPs.

3.8.1 BELLMAN EQUATIONS

Recall that in this model we do not have goals, and there are action rewards instead of costs. The Bellman equations are straightforward modifications of Equation 3.3:

$$
\begin{aligned}
V^*(s) &= \max_{a \in \mathcal{A}} Q^*(s, a) \\
Q^*(s, a) &= \sum_{s' \in \mathcal{S}} \mathcal{T}(s, a, s') \left[\mathcal{R}(s, a, s') + \gamma V^*(s') \right]
\end{aligned}
\tag{3.10}
$$

As in the case of SSP MDPs, they are just a restatement of the appropriate version of the Optimality Principle (Equation 2.5). Notice the absence of the base case checking if a state is a goal. The minimization is converted to a maximization, and future values are discounted by γ ($\gamma < 1$).

3.8.2 VALUE/POLICY ITERATION

VI and PI algorithms work as before, and moreover, the convergence results are stronger and the bounds tighter. Because there are no goals, the notion of a proper policy is not relevant. This implies that PI is guaranteed to converge *irrespective* of its initialization. Hence, it is a much more useful algorithm for these problems than for general SSP MDPs.

VI for infinite-horizon discounted-reward MDPs has stronger error bounds on the resulting policy upon reaching ϵ-consistency [242].

Theorem 3.24 The policy returned by Value Iteration that is run to ϵ-consistency satisfies the following inequality: $\forall s \in \mathcal{S},\ |V^*(s) - V^\pi(s)| < \frac{2\epsilon\gamma}{1-\gamma}$.

Additionally, we have tighter upper bounds on the number of iterations required in Value Iteration. To obtain an optimal policy, VI requires a number of iterations polynomial in $|\mathcal{S}|$, $|\mathcal{A}|$, and $\frac{1}{1-\gamma}$ [157; 236]. Such bounds have not yet been discovered for general SSP MDPs. For a detailed discussion of the theoretical properties of these algorithms for discounted reward MDPs, please refer to [197].

3.8.3 PRIORITIZED AND PARTITIONED ALGORITHMS

The original PS and Generalized PS algorithms are fully applicable for reward-maximization MDPs. However, extensions such as Improved PS, Focused Dynamic Programming, and Backward VI are designed specifically for goal-oriented cost-minimization settings. Analogous priority metrics are proposed for maximization settings also. For instance, a priority metric [246] that obtains good performance (as well as good cache performance) is:

$$\begin{aligned} priority(s) \quad &\leftarrow \quad Res^V(s) + V(s) \quad (\text{if } Res^V(s) > \epsilon) \\ &\leftarrow \quad 0 \quad \text{otherwise} \end{aligned}$$

This metric incorporates insights similar to Improved PS — higher-value states are more important, since from these the values will propagate to other lower reward states. Similar to IPS, it proposes a specific tradeoff between the residual and the value of a state — its performance is unpredictable and works great on some domains and not so on others.

Similarly, other priority-queue free algorithms have been explored for reward scenarios. An example is pre-sorting all states based on the maximum instantaneous reward achievable in a state and keeping the sorted order fixed throughout the computation. But, instead of backing all states, update only the predecessors of states where the residual was greater than ϵ in the previous iteration [68].

Partitioning-based algorithms extend easily to these approaches, and only small modifications, if at all, are needed to adapt them.

3.9 FINITE-HORIZON MDPS

Finite-Horizon MDPs are MDPs where the agent is allowed to act and gather rewards for a maximum time horizon T_{max} (see Definition 2.10). This is a special case of SSP MDPs, since any state after the horizon T_{max} is a trivial absorbing goal state.

Compared to previous MDPs, there is an important difference when defining the Bellman equations for finite-horizon problems. The agent's action depends not only on the world state s, but, also on the current decision epoch. As an example, if we have more time at hand, we could take a series of actions to gather a larger reward; however, if only a limited time is available, the agent may find a lower reward achievable in fewer actions to be preferable. We define $V^*(s, t)$ as the value of an *augmented state*, with agent at world state s at decision-epoch t, as follows:

$$
\begin{aligned}
V^*(s, t) &= 0 && (\text{if } t > T_{max}) \\
V^*(s, t) &= \max_{a \in \mathcal{A}} Q^*((s, t), a) && (t \le T_{max}) \\
Q^*((s, t), a) &= \sum_{s' \in \mathcal{S}} \mathcal{T}(s, a, s') \left[\mathcal{R}(s, a, s') + V^*(s', t+1) \right] && (3.11)
\end{aligned}
$$

The search space for Finite-Horizon MDPs blows up due to this additional horizon term. Algorithms, on the other hand, are much simpler and much more efficient (for similar-sized search spaces). This is because they exploit the observation that all finite-horizon problems have acyclic spaces – the agent can never return to the same augmented state, since time is always increasing.

Recall the discussion on acyclic MDPs in the example of Figure 3.1(a) and the optimal backup order Theorem 3.22. Acyclic MDPs do not require an iterative algorithm for solving MDPs. There is an optimal backup order that starts from the augmented states with maximum horizon value and incrementally computes the values of states with lower horizons. This can be achieved in a single pass updating the full layer of augmented states before moving to the previous layer. This strategy works for both value iteration and policy evaluation. Thus, VI, PI, and extensions under this backup order are guaranteed to terminate with *optimal* values (not just ϵ-consistent) in a single value update per augmented state.

3.10   MDPS WITH DEAD ENDS

According to Definition 2.20, SSP MDPs assume the existence of a complete proper policy, i.e., they assume that there exists a way for the agent to achieve a goal with probability 1 from *each* state. This assumption effectively prevents SSP MDPs from modeling situations in which the agent loses any chance of reaching the goal. For instance, with a small probability, the Mars rover could fall into a large crater while navigating and thereby become unable to achieve its scientific objectives. In general, scenarios involving catastrophic failures are quite common in the real world. Accidents, missing flight connections, going bankrupt are examples of such failures in navigation, travel planning, and economic planning domains respectively. In goal-oriented MDPs, consequences of such failures correspond to *dead ends*, states from which reaching the goal is impossible:

Definition 3.25   **Dead-End State.** A *dead-end state*, or *dead end*, is a state $s \in \mathcal{S}$ such that no policy can reach the goal from s in any number of time steps, i.e., for all policies π and any number of time steps t, $P_t^\pi(s, \mathcal{G}) = 0$.

SSP MDPs' inability to model domains with dead ends severely limits their applicability, so it is only natural to try to extend their solution techniques to handle these states. What happens if we allow dead-end states and use an algorithm such as VI as is? It is easy to see that $V^*(s)$ for a dead-end state s (Equation 3.3) is not well-defined – it diverges to infinity. Therefore, the algorithm will never converge.

3.10.1   FINITE-PENALTY SSP MDPS WITH DEAD-ENDS

One way to incorporate dead ends into SSP MDPs while preserving the properties of existing solution techniques for these problems is to assume that entering a dead end, while highly undesirable, has a *finite* price. For instance, suppose the agent buys an expensive ticket for a concert of a favorite band in another city, but remembers about it only on the day of the event. Getting to the concert venue requires a flight, either by hiring a business jet or by a regular airline with a layover. The first option is very expensive but almost guarantees making the concert on time. The second is much cheaper but, since the concert is so soon, missing the connection, a somewhat less probable outcome, means missing the concert. Nonetheless, the cost of missing the concert is still finite (say, the price of the ticket plus some finite value of missed enjoyment), so a rational agent would probably choose to travel with a regular airline.

To implement the finite-price assumption, we could assign a positive penalty $\mathcal{P}$ for visiting a dead end and no other change for other states. The semantics of this MDP would be that the agent pays $\mathcal{P}$ when encountering a dead end, and the process stops. However, this straightforward modification to SSP MDPs cannot be directly operationalized, since the set of dead ends is not known *a priori* and needs to be inferred while planning.

Moreover, this semantics also has a caveat — it may cause non-dead-end states that lie on potential paths to a dead end to have higher costs than dead ends themselves. For instance, imagine a state s whose only action (cost $\epsilon(\mathcal{P} + 1)$) leads with probability $1 - \epsilon$ to a dead end and with probability ϵ to a goal. A simple calculation shows that $V^*(s) = \mathcal{P} + \epsilon$, even though reaching a goal from s is possible. This is definitely undesirable. Moreover, notice that this semantic paradox cannot be resolved just by increasing the penalty $\mathcal{P}$, since the optimal value of state s will always be ϵ higher than the optimal value of a true dead end.

Therefore, we change the semantics of the finite-penalty model as follows. Whenever the agent reaches *any* state with the expected cost of reaching a goal equaling $\mathcal{P}$ or greater, the agent simply pays the penalty $\mathcal{P}$ and "gives up," i.e., the process stops. The benefit of putting a "cap" on all states' values in this manner is that the value function of a state under any policy becomes bounded, and we avoid the semantic pitfall above.

Intuitively, this setting describes scenarios where the agent can put a price on how desirable reaching a goal is. For instance, in the example involving a concert in another city, paying the penalty corresponds to deciding not to go to the concert, i.e., foregoing the pleasure it would have derived from attending the performance. We can define a class of MDPs describing such settings as follows [136]:

Definition 3.26   Finite-Penalty Stochastic Shortest-Path MDP with Dead Ends. A *finite-penalty stochastic shortest-path MDP with dead-ends (fSSPDE)* is a tuple $\langle \mathcal{S}, \mathcal{A}, \mathcal{T}, \mathcal{C}, \mathcal{G}, \mathcal{P} \rangle$, where $\mathcal{S}, \mathcal{A}, \mathcal{T}, \mathcal{C}$, and $\mathcal{G}$ have their usual meanings as in Definition 2.17 and $\mathcal{P} \in \mathbb{R}^+$ denotes the penalty incurred when an agent decides to abort the process in a non-goal state, under one condition:

- For every improper stationary deterministic Markovian policy π, for every $s \in S$ where π is improper, the value of π at s under the expected linear additive utility without the possibility of stopping the process by paying the penalty is infinite.

The objective of an agent in an fSSPDE MDP is to reach a goal with the minimum expected cost, including the possible penalty, *under the assumption that the agent can stop the process by paying the penalty.*

A natural question is: why does the definition need to make sure that improper policies have infinite costs under the assumption that the process cannnot be stopped, considering that this is not the criterion we are actually trying to optimize? The reason is that without this condition, under some policies that do not reach the goal, staying in a dead end could have only a finite cost and look so attractive that the agent would prefer going to that dead end rather than continuing towards a goal.

We can write Bellman equations for fSSPDE MDPs in the case where the set of all dead ends (D) is known *a priori*:

$$
\begin{aligned}
V^*(s) &= 0 && \text{(if } s \in \mathcal{G}) \\
&= \mathcal{P} && \text{(if } s \in D) \\
&= \min\left(\mathcal{P}, \min_{a \in \mathcal{A}} \sum_{s' \in \mathcal{S}} \mathcal{T}(s, a, s') \left[\mathcal{C}(s, a, s') + V^*(s')\right]\right) && \text{(otherwise)}
\end{aligned}
\tag{3.12}
$$

Since in reality the set of dead ends is not known, we cannot directly solve for these equations. However, there is a transformation enabling us to convert an fSSPDE MDP into an SSP. We can augment the action set $\mathcal{A}$ of an fSSPDE MDP with a special action a_{stop} that causes a transition to a goal state with probability 1, but at the cost of $\mathcal{P}$. This new MDP is an SSP, since reaching a goal with certainty is now possible from every state. The optimization criteria of *fSSPDE* and *SSP* yield the same set of optimal policies. We can also show that every SSP MDP can be cast as an fSSPDE MDP. In other words, we can show the following.

Theorem 3.27 $SSP = fSSPDE$ [136].

This suggests that all algorithms discussed in this chapter directly apply to fSSPDE problems. However, more experiments are needed to understand the efficiency hit due to dead ends and these algorithms' dependence on the value of $\mathcal{P}$. The algorithms will likely benefit if the set of dead ends could be pre-identified in advance [136].

To conclude, we remark that there are scenarios in which deadends need to be avoided at all costs, i.e., the dead-end penalty is *infinite*. The analysis of this case is rather complicated and we touch upon it only briefly when discussing iSSPUDE MDPs (Section 4.7.2).

CHAPTER 4

Heuristic Search Algorithms

4.1 HEURISTIC SEARCH AND SSP MDPS

The methods we explored in the previous chapter have a serious practical drawback — the amount of memory they require is proportional to the MDP's state space size. Ultimately, the reason for this is that they compute a complete policy, i.e., an optimal action for each state of an MDP.

For factored MDPs, the size of the state space (and hence of a complete policy) is exponential in the MDP description size, limiting the scalability of algorithms such as VI and PI. For instance, a factored MDP with 50 state variables has at least 2^{50} states. Even if each state could be represented with 50 bits, the total amount of memory required to represent such an MDP's complete policy (let alone the time to compute it!) would be astronomical. In the meantime, for many probabilistic planning problems we have two pieces of information that, as we will see in this chapter, can help us drastically reduce the amount of computational resources for solving them.

The first of these pieces of information is the MDP's initial state, denoted as s_0. In the Mars rover example, s_0 may be the configuration of the rover's systems (the amount of energy, the positions of its manipulators, etc.) and its location at the start of the mission. While computing a policy that reaches the goal from states other than s_0 may be useful, for many of the states doing so is unnecessary since they are not reachable from the initial state. In other words, if the initial state is known, it makes sense to compute only a *partial policy* π_{s_0} *closed w.r.t.* s_0.

Definition 4.1   Partial Policy. A stationary deterministic policy $\pi : \mathcal{S}' \to \mathcal{A}$ is *partial* if its domain is some set $\mathcal{S}' \subseteq \mathcal{S}$.

Definition 4.2   Policy Closed with Respect to State s. A stationary deterministic partial policy $\pi_s : \mathcal{S}' \to \mathcal{A}$ is *closed with respect to state s* if any state s' reachable by π_s from s is contained in $\mathcal{S}'$, the domain of π_s.

Put differently, a policy closed w.r.t. a state s must specify an action for any state s' that can be reached via that policy from s. Such a policy can be viewed as a restriction of a complete policy π to a set of states $\mathcal{S}'$ iteratively constructed as follows:

- Let $\mathcal{S}' = \{s\}$.

- Add to $\mathcal{S}'$ all the states that π can reach from the states in $\mathcal{S}'$ in one time step.

- Repeat the above step until $\mathcal{S}'$ does not change.

For many MDPs, a policy closed w.r.t. the initial state may exclude large parts of the state space, thereby requiring less memory to store. In fact, $\pi_{s_0}^*$, an *optimal* such policy, is typically even more compact, since its domain $\mathcal{S}'$ usually does not include all the states that are theoretically reachable from s_0 in *some* way. The advantages of computing a closed partial policy have prompted research into efficient solution techniques for a variant of stochastic shortest path problems where the initial state is assumed to be known. Below, we adapt the definition of SSP MDPs and concepts relevant to it to account for the knowledge of the initial state.

Definition 4.3   **Stochastic Shortest-Path MDP with an Initial State.** *(Strong definition.)* An *SSP MDP with an initial state*, denoted as SSP$_{s_0}$ MDP, is a tuple $\langle \mathcal{S}, \mathcal{A}, \mathcal{T}, \mathcal{C}, \mathcal{G}, s_0 \rangle$, where $\mathcal{S}, \mathcal{A}, \mathcal{T}, \mathcal{C},$ and $\mathcal{G}$ are as in the strong SSP MDP definition (2.20) and satisfy that definition's conditions, and $s_0 \in \mathcal{S}$ is the initial state, where the execution of any policy for this MDP starts.

The weak SSP MDP definition (2.17) can be modified analogously.

An optimal solution that we will be interested in finding when the initial state s_0 is known is $\pi_{s_0}^*$, a partial policy closed with respect to s_0 that minimizes the expected cost of reaching the goal *from s_0*, i.e., such that $V^*(s_0) \leq V^{\pi'}(s_0)$ for any other policy π'.

The second piece of information often available when modeling a problem as an MDP, and one that will help us find $\pi_{s_0}^*$ efficiently, is a *heuristic function*, or *heuristic*, for short. From an intuitive point of view, a heuristic is just some prior knowledge about the problem that lets us assess the quality of different states in the MDP. For instance, we may know that a state in which the Mars rover is stuck in a pit is "bad," because once the rover is stuck in the pit, cannot continue with its mission. Heuristics formalize such prior beliefs by associating a value with each state that measures how desirable a state is relative to others. These values can be computed automatically from the MDP description or simply estimated and hard-coded by an expert.

Definition 4.4   **Heuristic Function.** A *heuristic function* h is a value function that provides state values when an MDP solution algorithm inspects them for the first time.

How can a heuristic help us solve an SSP$_{s_0}$ MDP? Recall that our objective when solving such an MDP is to find an expected-cost-minimizing policy $\pi_{s_0}^*$. An obvious way to do that is to run a graph traversal algorithm to find all states reachable from s_0 and run VI or PI on them. A heuristic allows us to do better than that. Being optimal, $\pi_{s_0}^*$ will try to avoid states with a high expected cost of reaching the goal as much as possible. As a result, many such states will not be reachable by $\pi_{s_0}^*$ from s_0. Therefore, we would like to avoid spending computational resources on analyzing these states while constructing $\pi_{s_0}^*$. Intuitively, this is exactly what a heuristic helps us achieve. It makes some states look unattractive early on, encouraging their analysis only if, after several Bellman backup updates, other states start looking even less attractive. Thus, heuristics save computation time and space by concentrating search for $\pi_{s_0}^*$ on the more promising parts of the state space.

For SSP_{s_0} MDPs, the theory of heuristic search is relatively well-developed. In this chapter, we assume all MDPs to be of this type. We now examine algorithms that use a heuristic to efficiently find an optimal partial policy closed w.r.t. the initial state for such MDPs.

4.2   FIND-AND-REVISE: A SCHEMA FOR HEURISTIC SEARCH

In the previous section we gave a high-level idea of why heuristic search can be efficient, but have not explained how exactly we can employ heuristics for solving MDPs. In the remainder of this chapter we will see many algorithms for doing so, and all of them are instances of a general schema called FIND-and-REVISE (Algorithm 4.1) [31]. To analyze it, we need several new concepts.

Throughout the analysis, we will view the connectivity structure of an MDP M's state space as a *directed hypergraph* G_S which we call the *connectivity graph of M*. A directed hypergraph generalizes the notion of a regular graph by allowing each *hyperedge*, or *k-connector*, to have one source but several destinations. In the case of an MDP, the corresponding hypergraph G_S has S as the set of vertices, and for each state s and action a pair has a k-connector whose source is s and whose destinations are all states s' s.t. $\mathcal{T}(s, a, s') > 0$. In other words, it has a k-connector for linking each state via an action to the state's possible successors under that action. We will need several notions based on the hypergraph concept.

Definition 4.5   Reachability. A state s_n is *reachable* from s_1 in G_S if there is a sequence of states and actions $s_1, a_1, s_2, \ldots, s_{n-1}, a_{n-1}, s_n$, where for each i, $1 \leq i \leq n-1$, the node for s_i is the source of the k-connector for action a_i and s_{i+1} is one of its destinations.

Definition 4.6   The Transition Graph of an MDP Rooted at a State. *The transition graph of an MDP rooted at state s* is G_s, a subgraph of the MDP's connectivity graph G_S. Its vertices are s and only those states s' that are reachable from s in G_S. Its hyperedges are only those k-connectors that originate at s or at some state reachable from s in G_S.

When dealing with MDPs with an initial state, we will mostly refer to the transition graph G_{s_0} rooted at the initial state. This hypergraph includes only states reachable via some sequence of action outcomes from s_0. Historically, MDP transition graphs are also known as *AND-OR graphs*.

Definition 4.7   The Transition Graph of a Policy. *The transition graph of a partial deterministic Markovian policy* $\pi_s : S' \to A$ is a subgraph of the MDP's connectivity graph G_S that contains only the states in S' and, for each state $s \in S'$, only the k-connector for the action a s.t. $\pi(s) = a$.

Definition 4.8   The Greedy Graph of a Value Function Rooted at a State. *The greedy graph* G_s^V *of value function V rooted at state s* is the union of transition graphs of all policies π_s^V greedy w.r.t. V and closed w.r.t. s.

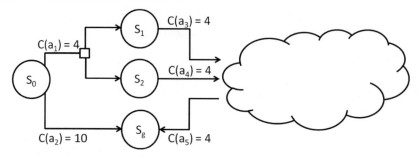

Figure 4.1: MDP showing a possible impact of a heuristic on the efficiency of policy computation via **FIND-and-REVISE**.

That is, G_s^V contains all states that can be reached via *some* π_s^V from s. As with general transition graphs, we will mostly be interested in greedy graphs of value functions rooted at the initial state s_0.

As the final pieces of terminology before we proceed, recall that the residual $Res^V(s)$ (Definition 3.16) is the magnitude of change in the value of a state as a result of applying a Bellman backup to value function V. A state s is called ϵ-consistent w.r.t. V if $Res^V(s) < \epsilon$ (Definition 3.17) and ϵ-inconsistent otherwise.

Algorithm 4.1: FIND-and-REVISE

1 Start with a heuristic value function $V \leftarrow h$
2 **while** V's greedy graph $G_{s_0}^V$ contains a state s with $Res^V(s) > \epsilon$ **do**
3 $\quad$ FIND a state s in $G_{s_0}^V$ with $Res^V(s) > \epsilon$
4 $\quad$ REVISE $V(s)$ with a Bellman backup
5 **end**
6 return a π^V

The idea of **FIND-and-REVISE** is quite simple. It iteratively searches the greedy graph of the current value function for an ϵ-inconsistent state and updates the value of that state and possibly of a few others with a finite number of Bellman backups. This typically changes the greedy graph, and the cycle repeats. **FIND-and-REVISE** can be viewed as a Prioritized Value Iteration scheme (Section 3.5) over the MDP's transition graph rooted at s_0 — in every iteration it assigns a higher priority to every state in $G_{s_0}^V$ than to any state outside of this greedy graph. Moreover, the FIND step can use values of states within the greedy graph for further prioritization. Recall that Prioritized Value Iteration is itself a kind of Asynchronous Value Iteration (Section 3.4.4), so **FIND-and-REVISE** is a member of this more general class of approaches as well.

Crucially, the greedy graph that **FIND-and-REVISE** starts with is induced by some heuristic function h. To demonstrate the difference h can make on the number of states a **FIND-and-REVISE**-like algorithm may have to store, we present the following example.

Example: Consider the transition graph G_{s_0} of the SSP_{s_0} MDP in Figure 4.1. This MDP has many states, four of which (the initial state s_0, the goal state s_g, and two other states s_1 and s_2) are shown, while the rest are denoted by the cloud. Action costs are shown for a subset of the MDP's actions; assume that costs are nonnegative for actions in the cloud. At least one of the actions, a_1, has several probabilistic effects, whose probabilities do not matter for this example and are omitted. Note that $\pi_{s_0}^*$ for this MDP is unique and involves taking action a_2 from s_0 straight to the goal; thus, $V^*(s_0) = 10$. All other policies involve actions a_1, a_5, and a_3 or a_4. Therefore, the cost of reaching the goal from s_0 using them is at least $4 \cdot 3 = 12 > 10$, making these policies suboptimal.

Now, the transition graph G_{s_0} of the MDP in Figure 4.1 can be arbitrarily large, depending on the number of states in the cloud. This is the largest set of states an MDP solution algorithm may *have to* store while searching for $\pi_{s_0}^*$. Compare G_{s_0} to $G_{s_0}^{V^*}$, the greedy graph of the optimal value function. $G_{s_0}^{V^*}$ contains only the states visited by $\pi_{s_0}^*$, i.e., s_0 and s_g, as established above. This is the very smallest set of states we can hope to explore while looking for $\pi_{s_0}^*$. Finally, consider $G_{s_0}^h$ for a heuristic value function h that assigns $h(s_g) = 0$ and, for instance, $h(s_1) = h(s_2) = 7$. These values would induce $Q^h(s_0, a_1) = 4 + 7 = 11 > Q^h(s_0, a_2) = 10$, making a_2 more preferable in s_0 and thus immediately helping discover the optimal policy. Thus, $G_{s_0}^h$ consists of s_0 and s_g, and starting FIND-and-REVISE from such an h allows FIND-and-REVISE to evaluate only four states before finding $\pi_{s_0}^*$. Contrast this with an algorithm such as VI, which, even initialized with a very good h, will still necessarily visit the entire transition graph rooted at the initial state, G_{s_0}. As this example shows, FIND-and-REVISE in combination with a good heuristic can make finding $\pi_{s_0}^*$ arbitrarily more efficient than via VI or PI!   □

The fact that FIND-and-REVISE may never touch some of the states, as in the above example, might seem alarming. After all, Prioritized Value Iteration algorithms in general fail to find an optimal policy if they starve some states, and FIND-and-REVISE appears to be doing exactly that. As it turns out, however, if the FIND-and-REVISE's FIND procedure is *systematic* and the heuristic function FIND-and-REVISE is using is *admissible*, FIND-and-REVISE is guaranteed to converge to an optimal solution for a sufficiently small ϵ.

Definition 4.9   Systematicity. Let $K_i(s)$ be the total number of FIND-and-REVISE iterations that state s has spent in the greedy graph rooted at s_0 with $Res^V(s) > \epsilon$ between iterations i and $i + K$. The FIND-and-REVISE's FIND procedure is called *systematic* if for all $i > 1$ and for all $s \in S$ the probability that FIND will choose s for the REVISE step at least once after iteration i approaches 1 as $K_i(s)$ goes to ∞.

This technical definition has an intuitive meaning — a FIND procedure is systematic (i.e., searches the greedy graph systematically) if it does not starve any of the states possibly relevant to finding the optimal solution. Such a FIND procedure will not allow an ϵ-inconsistent state to stay in the greedy graph forever without its value being revised. At the same time, it may ignore states that at some point leave the greedy graph for good.

Definition 4.10   Heuristic Admissibility. A heuristic h is *admissible* if for all states s in the transition graph G_{s_0}, $h(s) \leq V^*(s)$. Otherwise, the heuristic is called *inadmissible*.

Theorem 4.11   For an SSP_{s_0} MDP, if FIND-and-REVISE is initialized with an admissible heuristic, terminated when all the states in the value function's greedy graph rooted at s_0 are ϵ-consistent, and its FIND procedure is systematic, the value function computed by FIND-and-REVISE approaches the optimal value function over all states reachable from s_0 by an optimal policy as ϵ goes to 0 [31].

Theorem 4.11 contains a small caveat. As with VI, although *in the limit* a vanishingly small residual implies optimality, for finite values of ϵ FIND-and-REVISE can return a significantly suboptimal policy. Nonetheless, in practice this rarely happens. In this book we will assume the chosen ϵ to be small enough to let FIND-and-REVISE halt with an optimal policy.

We will examine several techniques for computing admissible heuristics automatically in Section 4.6, and until then will simply assume the SSP MDP heuristic we are given to be admissible, i.e., to be a lower bound on V^*. This assumption has a very important implication. Recall from Chapter 3 that Bellman backups on SSP MDPs are monotonic (Theorem 3.20); in particular, if for some V, $V \leq V^*$, then after V is updated with a Bellman backup it is still true that $V \leq V^*$. Therefore, if FIND-and-REVISE starts with an admissible heuristic, after every state value update the resulting value function is still a lower bound on V^*. In effect, FIND-and-REVISE generates a sequence of value functions that approaches V^* from below for all states visited by $\pi_{s_0}^*$. To stress this fact, we will refer to the value function V FIND-and-REVISE maintains as V_l in the rest of this chapter (l stands for "lower bound").

FIND-and-REVISE's pseudocode intentionally leaves the FIND and REVISE procedures unspecified — it is in their implementations that various heuristic search algorithms differ from each other. Their REVISE methods tend to resemble one another, as they are all based on the Bellman backup operator. Their FIND methods, however, can be vastly distinct. An obvious approach to finding ϵ-inconsistent states is via a simple systematic search strategy such as depth-first search. Indeed, depth-first search is used in this manner by several proposed FIND-and-REVISE algorithms, e.g., HDP [31] and LDFS [33]. Employing depth-first search for the purpose of identifying weakly explored states also echoes similar approaches in solving games, non-deterministic planning problems, and other related fields [33]. However, in tackling MDPs this strategy is usually outperformed by more sophisticated search methods, which we are about to explore.

4.3   LAO* AND EXTENSIONS

Historically, the first FIND-and-REVISE algorithm that could deal with arbitrary SSP_{s_0} MDPs was LAO* [105]. We start our presentation of heuristic search techniques with this algorithm as well and then describe its derivatives.

4.3.1   LAO*

The basic approach of LAO* is to look for states with high residuals only in a restricted region of the current value function's greedy policy graph (AND-OR graph, in LAO* terminology) reachable from the initial state, called the *solution graph*. Besides the solution graph, LAO* also keeps track of a superset of the solution graph's nodes called the *envelope*, which contains all the states that have ever been part of the solution graph in the past. By updating values of states in the envelope, LAO* constructs a new greedy policy rooted at s_0 (and the solution graph corresponding to it) over the envelope's states. Sometimes, this policy, "attracted" by the low heuristic values of states in the unexplored area of state space, takes an action leading the agent to a state beyond the envelope. This causes LAO* to gradually expand and modify the envelope. At some point, LAO* ends up with a greedy policy over the envelope that never exits the envelope and that cannot be improved. This policy is optimal.

Pseudocode for LAO* is presented in Algorithm 4.2. Starting with the initial state, it gradually expands $\hat{G}_{s_0}$, the *explicit graph* of the envelope, by adding new states to it. $\hat{G}_{s_0}$ represents the connectivity structure of the states in the envelope and is a subgraph of the transition graph G_{s_0} explored by LAO* so far. Thus, at each iteration of LAO*, the current envelope is simply the set of $\hat{G}_{s_0}$'s nodes. This set can be subdivided into the set of the envelope's *fringe* states F and the set of *interior* states I. Each fringe state has at least one successor under some action that lies outside of $\hat{G}_{s_0}$. In contrast, all successors of all interior states belong to $\hat{G}_{s_0}$.

Using $\hat{G}_{s_0}$, LAO* also maintains the solution graph $\hat{G}_{s_0}^{V_l}$ — a subgraph of $\hat{G}_{s_0}$ representing the current best partial policy over the current envelope, rooted at s_0. That is, $\hat{G}_{s_0}^{V_l}$ is a graph of all states in $\hat{G}_{s_0}$ that a policy greedy w.r.t V_l can reach from the initial state.

When LAO* starts running, $\hat{G}_{s_0}$ contains only s_0, and this state is $\hat{G}_{s_0}$'s fringe state (lines 4-6 of Algorithm 4.2). In subsequent iterations, LAO* chooses a fringe state s of $\hat{G}_{s_0}$ that also belongs to the solution graph $\hat{G}_{s_0}^{V_l}$ and expands it (lines 10-14), i.e., adds all the successors of s under all possible actions to $\hat{G}_{s_0}$, except for those that are in $\hat{G}_{s_0}$ already. Such a state can be chosen arbitrarily among all fringe states reachable from s_0 in the solution graph. However, a clever way of choosing the state to expand can greatly improve LAO*'s performance. Once expanded, a fringe state becomes an internal state. Next, using Bellman backups, LAO* updates the values of the expanded state and any state of $\hat{G}_{s_0}$ from which the expanded state can be reached within $\hat{G}_{s_0}$ via a greedy policy (lines 16-17). This set of states, denoted as Z in the pseudocode, is the set of all states in the envelope whose values could have changed as a result of envelope expansion. The state value updates over Z can be performed by running either PI or VI. The pseudocode in Algorithm 4.2 uses PI; however, LAO* can also be implemented using VI, and we refer the reader to the original paper for the details of that version [105]. It is at this step that the heuristic comes into play — the heuristic initializes values of the successors of the newly expanded state. PI/VI propagates these values into the interior of $\hat{G}_{s_0}$, potentially letting them affect the current greedy policy.

Finally, LAO* rebuilds $\hat{G}_{s_0}^{V_l}$ (line 18), the solution graph rooted at s_0, which could have changed because V_l was updated in the previous step. The algorithm then proceeds to the next

---

Algorithm 4.2: LAO*

1 // *Initialize V_l, the fringe set F, the interior set I, the explicit graph of the envelope $\hat{G}_{s_0}$,*
2 // *and the solution graph $\hat{G}_{s_0}^{V_l}$ over states in the explicit graph.*
3 $V_l \leftarrow h$
4 $F \leftarrow \{s_0\}$
5 $I \leftarrow \emptyset$
6 $\hat{G}_{s_0} \leftarrow \{\text{nodes: } I \cup F, \text{ hyperedges: } \emptyset\}$
7 $\hat{G}_{s_0}^{V_l} \leftarrow \{\text{nodes: } \{s_0\}, \text{ hyperedges: } \emptyset\}$
8 **while** $F \cap \hat{G}_{s_0}^{V_l}$ *has some non-goal states* **do**
9 $\quad$ // *Expand a fringe state of the best partial policy*
10 $\quad$ $s \leftarrow$ some non-goal state in $F \cap \hat{G}_{s_0}^{V_l}$
11 $\quad$ $F \leftarrow F \setminus \{s\}$
12 $\quad$ $F \leftarrow F \cup \{\text{all successors } s' \text{ of } s \text{ under all actions, } s' \notin I\}$
13 $\quad$ $I \leftarrow I \cup \{s\}$
14 $\quad$ $\hat{G}_{s_0} \leftarrow \{\text{nodes: } I \cup F, \text{ hyperedges: all actions in all states of } I\}$
15 $\quad$ // *Update state values and mark greedy actions*
16 $\quad$ $Z \leftarrow \{s$ and all states in $\hat{G}_{s_0}$ from which s can be reached via the current greedy policy$\}$
17 $\quad$ Run PI on Z until convergence to determine the current greedy actions in all states in Z
18 $\quad$ Rebuild $\hat{G}_{s_0}^{V_l}$ over states in $\hat{G}_{s_0}$
19 **end**
20 return $\pi_{s_0}^{V_l}$

---

iteration. This process continues until $\hat{G}_{s_0}^{V_l}$ has no more nonterminal fringe states, i.e., states that are not goals.

Viewed in terms of the FIND-and-REVISE framework, LAO* in essence lazily builds the transition and the greedy graph. In each iteration, it adds to both graphs' partial representations $\hat{G}_{s_0}$ and $\hat{G}_{s_0}^{V_l}$, respectively, by expanding a fringe node. LAO* then updates all the states of $\hat{G}_{s_0}$ whose values the expansion could have possibly changed, modifies $\hat{G}_{s_0}^{V_l}$ if necessary, and so on. Identifying the states whose values could have changed corresponds to the FIND-and-REVISE's FIND procedure, since the residuals at these states are likely to be greater than ϵ. Updating these states' values corresponds to the REVISE step.

The readers familiar with heuristic search in regular graphs, as exemplified by the A* algorithm [106], may have noticed some similarity in the operation of A* and LAO*. Both algorithms gradually construct a partial solution. A* does this in a regular graph, and its partial solutions are regions of the graph corresponding to *partial linear plans* originating at s_0. LAO* constructs a solution in a hypergraph, not in a regular graph. Accordingly, the partial solutions it deals with are regions of the hypergraph corresponding to *partial policies* rooted at s_0. Indeed, LAO* is a generalization of A*'s heuristic search idea to probabilistic planning problems. Each of these algorithms employs a *heuristic* to guide the modification of the current partial policy/plan, i.e., to *search* the space of solutions.

4.3.2   ILAO*

The most computationally expensive part of LAO* is running VI/PI on states whose values are affected by the envelope expansions. LAO* runs these algorithms to (near-)convergence after every fringe state expansion and ends up spending a lot of time on each such step. Other implementations are also possible, in which multiple states are expanded in each iteration or fewer backups are performed. All such implementations are guaranteed to converge as long as the FIND-and-REVISE conditions are met. The best balance between expansion and backups may be problem-dependent.

One such implementation, Improved LAO* (ILAO*) [105], differs from LAO* in three ways:

- During each $\hat{G}_{s_0}^{V_l}$ expansion attempt, instead of selecting one fringe state of the greedy partial policy for expansion, ILAO* expands all fringe states of that policy.

- Instead of running a full VI/PI, ILAO* backs up all states in $\hat{G}_{s_0}^{V_l}$ only once.

- The backups above are performed in a depth-first postorder traversal of $\hat{G}_{s_0}^{V_l}$.

ILAO* is implemented using a depth-first search routine starting from the initial state. At each interior state, the greedy successors are pushed onto the DFS stack. All non-goal fringe states are expanded, and their successor values are initialized by the heuristic. After all the expansions are done, the DFS stack contains all states in the current greedy graph. ILAO* now pops these states from the stack and performs one backup per state. This results in a postorder backup order where all successors are backed up before the parent.

These iterations are repeated until the greedy policy graph has no nonterminal states and the residual is less than ϵ. At termination, the greedy policy graph is returned as the solution. These optimizations mitigate LAO*'s inefficiencies by reducing the expensive operations (backups) and increasing the expansions in each iteration.

4.3.3   BLAO* AND RLAO*: EXPANDING THE REVERSE ENVELOPE

After LAO* starts running and until at least one goal state is included in the envelope, the envelope expansions are guided primarily by the heuristic. Depending on the peculiarities of the heuristic, "discovering" a goal state via envelope expansions from the initial state may take many iterations of LAO*'s main loop. Because of this, it may take a while before the choice of action in s_0 becomes informed by the signal from the goal, not the heuristic values at the fringe. For deterministic heuristic-search algorithms such as A*, one way to address this issue is to search from the goal states to the initial state, or both backward from the goal and forward from the initial state. Reverse LAO* (RLAO*) [58] and Bidirectional LAO* (BLAO*) [25] implement these ideas for MDPs.

RLAO* uses a reverse explicit graph rooted at the goal state(s) to compute a policy. Expanding a state s' in this graph means identifying all or some of the predecessor states s from which s' can be reached by executing a single action. RLAO* also maintains a best policy graph over the states in the reverse explicit graph, and grows the reverse explicit graph by adding to it in every iteration all the predecessors of all the states on the fringe of the best policy graph. Consequently, its performance is strongly dependent on the fan-in (i.e., the average number of states from which one can transition to a given state). It suffers noticeably when the fan-in is high.

BLAO* expands both a forward and a backward explicit graph. With every expansion of the forward explicit graph in the same way as LAO*, it also expands the reverse graph by adding to it a state s connected to a state s' already in the reverse graph via an action with the lowest Q-value. Intuitively, this amounts to adding the "best" predecessor to the reverse graph. Eventually, the forward and the reverse graphs overlap, establishing a "bridge" between the initial state and the goal. BLAO* sometimes converges several times faster than LAO*. It does not suffer from high fan-in as much as RLAO*, because it expands the reverse graph less aggressively.

4.3.4 AO*: HEURISTIC SEARCH FOR ACYCLIC MDPS

In the previous few subsections, we have described nearly all of LAO*'s derivatives; however, an exposition of LAO*-type approaches would be incomplete without also mentioning LAO*'s direct *predecessor*, the AO* [189] algorithm. As noted in passing, LAO* can be viewed as a generalization of the deterministic planning algorithm A* to probabilistic planning. AO* [189] is the "missing link" between the two. It is an extension of A* to *acyclic* MDPs, probabilistic planning problems that do not have cycles. In such problems, after an agent has visited some state s, its probability of returning to s via any sequence of actions is 0. We have already seen an SSP MDP subclass that has this property, the finite-horizon MDPs (Section 3.9). In finite-horizon MDPs, the agent transitions between states in an augmented state space. Since the current time step is part of the state specification and every action execution increments the time counter, going back to an already visited augmented state is impossible. Thus, AO* can be viewed as a specialization of LAO* to finite-horizon MDPs.

Structurally, the AO* and LAO* algorithms are very similar. Both iterate over two steps — expansion of the current best partial policy, and updating states' values in the current policy envelope to propagate heuristic values from the newly expanded state. In fact, lines 1-16 of LAO* (Algorithm 4.2) are identical for AO*. The only difference lies in these algorithms' methods of updating state values after adding new states to the policy envelope. Note that due to the potential presence of cycles in a policy graph of a general SSP MDP, updating state values in it requires an iterative dynamic programming approach, and LAO* uses PI or VI for this purpose. On the other hand, in an acyclic MDP the update procedure need not be so complicated. Instead, it can be performed in a bottom-up fashion in a single sweep; the heuristic values at the tips of the policy graph can be propagated to the graph's internal states by applying Bellman backups to the parents of the tips, then to the parents of the parents of the tips, and so on. This is exactly what AO* does in place of LAO*'s use of VI/PI to update state values. After the update step, starting from LAO*'s line 18, the pseudocode of AO* and LAO* coincides again.

Although AO* can handle only a strict subset of MDPs solvable by LAO*, it is much more efficient on this subset, because it can update all relevant states in a single sweep. One might think that VI and PI would converge after one sweep as well if launched on an acyclic graph, which would make LAO* as fast as AO* on acyclic MDPs. This is partly true, with one caveat — VI/PI would need to update states *in the "right" order*. That is, on acyclic MDPs, AO* is equivalent to LAO* with a special prioritization strategy (Section 3.5).

4.4   RTDP AND EXTENSIONS

An alternative to the LAO*-style gradual, systematic expansion of the solution graph is a scheme called Real-Time Dynamic Programming (RTDP) [12]. The motivation for RTDP comes from scenarios where an agent (e.g., a robot) is acting in the real world and only occasionally has time for planning between its actions. Such an agent needs a (preferably distributed) algorithm capable of quickly exploring various areas of the state space and coming up with a reasonable action to execute in the agent's current state before planning time has run out. Thanks to its sampling-based nature, RTDP satisfies these requirements, and improvements upon it that we will also examine endow the basic approach with other useful properties.

4.4.1   RTDP

At a high level, RTDP-based algorithms operate by simulating the current greedy policy to sample "paths," or *trajectories*, through the state space, and performing Bellman backups only on the states in those trajectories. These updates change the greedy policy and make way for further state value improvements.

The process of sampling a trajectory is called a *trial*. As shown in Algorithm 4.3, each trial consists of repeatedly selecting a greedy action a_{best} in the current state s (line 15), performing a Bellman backup on the value of s (line 16), and transitioning to a successor of s under a_{best} (line 17). A heuristic plays the same role in RTDP as in LAO* — it provides initial state values in order to guide action selection during early state space exploration.

Each sampling of a successor during a trial that does not result in a termination of the trial corresponds to a FIND operation in FIND-and-REVISE, as it identifies a possibly ϵ-inconsistent state to update next. Each Bellman backup maps to a REVISE instance.

The original RTDP version [12] has two related weaknesses, the main one being the lack of a principled termination condition. Although RTDP is guaranteed to converge asymptotically to V^* *over the states in the domain of an optimal policy* $\pi^*_{s_0}$, it does not provide any mechanisms to detect when it gets near the optimal value function or policy. The lack of a stopping criterion, although unfortunate, is not surprising. RTDP was designed for operating under time pressure, and would almost never have the luxury of planning for long enough to arrive at an optimal policy. In these circumstances, a convergence detection condition is not necessary.

Algorithm 4.3: RTDP

```
 1  RTDP(s_0)
 2  begin
 3  │   V_l ← h
 4  │   while there is time left do
 5  │   │   TRIAL(s_0)
 6  │   end
 7  │   return π*_{s_0}
 8  end
 9
10
11  TRIAL(s_0)
12  begin
13  │   s ← s_0
14  │   while s ∉ G do
15  │   │   a_{best} ← argmin_{a∈A} Q^{V_l}(s, a)
16  │   │   V_l(s) ← Q^{V_l}(s, a_{best})
17  │   │   s ← execute action a_{best} in s
18  │   end
19  end
```

The lack of convergence detection leads to RTDP's other drawback. As RTDP runs longer, V_l at many states starts to converge. Visiting these states again and again becomes a waste of resources, yet this is what RTDP keeps doing because it has no way of detecting convergence. An extension of RTDP we will look at next addresses both of these problems by endowing RTDP with a method of recognizing proximity to the optimal value function over relevant states.

4.4.2   LRTDP

Labeled RTDP (LRTDP) [32] works in largely the same way as RTDP, but also has a mechanism for identifying ϵ-consistent states and marking them as solved. To provide a theoretical basis for LRTDP's termination condition, we need the following definition:

Definition 4.12   Monotone Lower Bound Value Function. A value function V of an SSP MDP is a *monotone lower bound* on V^* if and only if for any $s \in \mathcal{S}$, $V(s) \leq \min_{a \in \mathcal{A}} Q^V(s, a)$.

The intuitive characteristic of a monotone lower bound is that such a value function can only grow as the result of a Bellman backup update. The monotonicity of an admissible heuristic has an important consequence for the convergence of RTDP.

**Theorem 4.13   **For RTDP initialized with a monotone lower bound, the current value function V is ϵ-consistent at state s and will remain ϵ-consistent at this state indefinitely if $Res^V(s) < \epsilon$ and $Res^V(s') < \epsilon$ for all descendants s' of s in the greedy graph G_s^V.

In other words, s is guaranteed to remain ϵ-consistent forever from the point when Bellman backups cannot change either $V(s)$ or the values of any of s's descendants in G_s^V by more than ϵ. Informally, the reason for this is that the value $V(s)$ of a state is determined solely by the values of its descendants in the (implicitly maintained) greedy graph G_s^V. Therefore, hypothetically, $V(s)$ can change by more than ϵ only under two circumstances — either if G_s^V changes, or if the values of some of s's descendants in G_s^V change by more than ϵ. As it turns out, in a FIND-and-REVISE algorithm started from a monotone admissible heuristic, the sole way to modify G_s^V is by updating a value of a state *within* G_s^V. Updating states outside G_s^V will never make them part of G_s^V because, by the monotonicity property, Bellman backups can only increase the values of states and therefore make them only less attractive and less eligible to be part of a greedy graph. This implies that there is actually just one way for RTDP to change $V(s)$ by more than ϵ — by changing a value of a descendant of s in G_s^V by more than ϵ. However, this would contradict the premise of the above theorem that all the descendants of s are already ϵ-consistent.

LRTDP (Algorithm 4.4) implements a mechanism for detecting convergence implied by the above theorem in its CHECK-SOLVED method (Algorithm 4.5). To verify that the value of a state s has stabilized, CHECK-SOLVED checks whether the residual at s or any of its descendants in $G_s^{V_l}$ exceeds ϵ (lines 10-12 of Algorithm 4.5). For this purpose, it keeps two stacks — *open*, with states still to be checked for ϵ-consistency, and *closed*, with already checked states. In every iteration it takes a state off the *open* stack (line 8), moves it onto *closed* (line 9), and sees whether the state's residual is greater than ϵ. If so, s cannot be labeled as solved yet (line 11). Otherwise, CHECK-SOLVED expands the state just examined (lines 14-18) in order to check this state's successors as well. In this way, all descendants of states s' in $G_s^{V_l}$ with $Res^{V_l}(s') < \epsilon$ eventually end up being examined by CHECK-SOLVED. If the residuals of all these states are smaller than ϵ, all of them, including s, are labeled solved (lines 22-24). Otherwise, those whose residuals are at least ϵ get updated with Bellman backups (lines 27-30).

LRTDP uses CHECK-SOLVED to label states it visits during the trials (lines 24-28 of Algorithm 4.4) and, most importantly, to terminate trials early once they end up at a labeled state (line 15). The labeling procedure makes LRTDP's convergence to an ϵ-consistent value function orders of magnitude faster than RTDP's [32]. Moreover, it makes LRTDP useful not only in real-time settings but for offline planning as well.

LRTDP's versatility, simplicity, and efficiency when equipped with a good heuristic make it a popular choice for many problems. Its derivative for finite-horizon problems, LR^{2}TDP [130], has been used as the main component of two competitive planners, Glutton [130; 204] and Gourmand [135]. At the same time, like LAO*, LRTDP still leaves room for improvement. We explore some of its variants next.

4.4.3   BRTDP, FRTDP, VPI-RTDP: ADDING AN UPPER BOUND

LRTDP has several areas for further improvement. When choosing the next state in a trial, LRTDP simply samples it from the transition function of the current greedy action. Instead, it could bias its selection toward poorly explored states, which could make convergence faster and more uniform.

Algorithm 4.4: LRTDP

```
 1  LRTDP(s₀, ε)
 2  begin
 3  │   Vₗ ← h
 4  │   while s₀ is not labeled solved do
 5  │   │   LRTDP-TRIAL(s₀, ε)
 6  │   end
 7  │   return π*_{s₀}
 8  end
 9
10
11  LRTDP-TRIAL(s₀, ε)
12  begin
13  │   visited ← empty stack
14  │   s ← s₀
15  │   while s is not labeled solved do
16  │   │   push s onto visited
17  │   │   if s ∈ 𝒢 then
18  │   │   │   break
19  │   │   end
20  │   │   a_best ← argmin_{a∈𝒜} Q^{Vₗ}(s, a)
21  │   │   Vₗ(s) ← Q^{Vₗ}(s, a_best)
22  │   │   s ← execute action a_best in s
23  │   end
24  │   while visited ≠ empty stack do
25  │   │   s ← pop the top of visited
26  │   │   if ¬CHECK-SOLVED(s, ε) then
27  │   │   │   break
28  │   │   end
29  │   end
30  end
```

That is, if viewed as a kind of Prioritized Value Iteration (Section 3.5), LRTDP uses just one of many possible prioritization strategies, and not necessarily the best one. In addition, LRTDP provides few quality guarantees if stopped *before* it converges and asked to return the current best policy. In fact, as already mentioned, a policy greedy w.r.t. a lower bound on V^* could be arbitrarily bad [176]. This is a serious drawback in domains such as robotics, where due to time pressure the agent (robot) may want to stop planning as soon as policy quality reaches a particular threshold.

The three RTDP variants we describe next address these issues by maintaining a monotone *upper* bound V_u on V^* in addition to a lower bound V_l. Similar to a monotone lower bound (Definition 4.12), a monotone upper bound is a value function $V_u \geq V^*$ s.t. applying a Bellman backup to it results in a closer approximation V_u' of V^*, i.e., $V_u' \leq V_u$. As with V_l, these algorithms

---

Algorithm 4.5: CHECK-SOLVED

---

1 CHECK-SOLVED(s_0', ϵ)
2 **begin**
3      $retVal \leftarrow true$
4      $open \leftarrow$ empty stack
5      $closed \leftarrow$ empty stack
6   push s_0' onto $open$
7   **while** $open \neq empty\ stack$ **do**
8          $s \leftarrow$ pop the top of $open$
9   push s onto $closed$
10   **if** $Res^{V_l}(s) > \epsilon$ **then**
11             $retVal \leftarrow false$
12   continue
13   **end**
14         $a_{best} \leftarrow \arg\min_{a \in \mathcal{A}} Q^{V_l}(s, a)$
15   **foreach** $s' \in \mathcal{S}$ s.t. $\mathcal{T}(s, a_{best}, s') > 0$ **do**
16   **if** s' is not labeled solved and $s' \notin open \cup closed$ **then**
17   push s onto $open$
18   **end**
19   **end**
20   **end**
21   **if** $retVal == true$ **then**
22   **foreach** $s \in closed$ **do**
23   label s solved
24   **end**
25   **end**
26   **else**
27   **while** $closed \neq empty\ stack$ **do**
28             $s \leftarrow$ pop the top of $closed$
29             $V_l(s) \leftarrow \min_{a \in A} Q^{V_l}(s, a)$
30   **end**
31   **end**
32   return $retVal$
33 **end**

---

initialize V_u with a heuristic (although a "strictly inadmissible" one) and update it with Bellman backups during the trials.

How can the knowledge of an upper bound help us improve upon LRTDP? To start with, the difference between the bounds at a given state, ($V_u(s) - V_l(s)$), is a natural measure of uncertainty about the state's value — roughly, the larger the gap, the greater the uncertainty. We can use this information to guide state space exploration toward states at which value function uncertainty is high. Furthermore, ($V_u(s) - V_l(s)) < \epsilon$ provides a convergence criterion that, unlike the labeling procedure of LRTDP, does not require recursive computation. Last but not least, a policy π greedy w.r.t. a monotone V_u has the following crucial property.

Theorem 4.14   Suppose V_u is a monotone upper bound on V^*. If π is a greedy policy w.r.t. V_u, then $V^\pi \leq V_u$ [176].

As a consequence, the expected cost of a greedy policy derived from V_u at any state s is no greater than $V_u(s)$, yielding a quality guarantee.

These observations have given rise to several RTDP-based algorithms: Bounded RTDP (BRTDP) [176], Focused RTDP (FRTDP) [219], and Value of Perfect Information RTDP (VPI-RTDP) [210]. All of them explore the state space in a trial-based fashion, like other RTDP-family approaches. All of them maintain an upper and a lower bound on V^*, updating both with Bellman backups. All of them select actions during trials greedily w.r.t. V_l, but return a policy greedy w.r.t. V_u upon termination. Their convergence detection mechanisms are also similar, terminating the algorithm when the gap $(V_u(s) - V_l(s))$ becomes small for all states reachable from s_0 by the current best policy. What they primarily differ in are strategies for biasing state space exploration.

Bounded RTDP (BRTDP) [176] focuses its trajectories on states with large gaps between the lower and the upper bound, i.e., those where the value function is "poorly understood." It does so by sampling the next state s' in a trajectory with a probability proportional to $\mathcal{T}(s, a, s')(V_u(s') - V_l(s'))$, where s is the current state and a is a greedy action in s with respect to V_l.

FRTDP [219], instead of sampling the next state the way BRTDP does, prioritizes the successors under a V_l-greedy action in a state s according to a scoring function, and deterministically selects the highest-ranked one. The scoring function is a product of the uncertainty in the state's value as reflected by $(V_u(s') - V_l(s'))$ and a measure of the fraction of time steps the current best policy is expected to spend in s' before entering a goal state. Thus, a state's score is an approximation of the expected "benefit" from updating the state.

VPI-RTDP [210] directs state space exploration where the *policy*, as opposed to the value function, is furthest from convergence. In this way, it improves upon FRTDP and BRTDP, which may waste computation on states with an unconverged value function but converged policy. When choosing a successor s' of s under action a, it first checks whether $(V_u(s') - V_l(s'))$ is smaller than some threshold β for all possible successors. If $(V_u(s') - V_l(s')) > \beta$ for all successors s', the value function at all the successors is far from convergence, so exploring any of them will likely result in a significant change in policy. Therefore, in this case VPI-RTDP samples a successor the same way as BRTDP does. However, if for some successors the gap is small, for each successor s' VPI-RTDP calculates $VPI(s')$, an approximation of the value of perfect information about the optimal policy that can be obtained by updating s'. Roughly, $VPI(s')$ measures the expected change in policy due to updating s'. If $VPI(s')$ is non-zero for at least one successor of s, VPI-RTDP samples a successor with probability proportionate to $VPI(s')$. Otherwise, it falls back to the BRTDP method of successor selection.

4.5   HEURISTICS AND TRANSITION GRAPH PRUNING

Heuristic search algorithms can produce an optimal MDP solution relatively efficiently *if* a good admissible heuristic is available, but give no clue how to actually obtain such a heuristic. In this section, we study several ideas for computing admissible heuristics, as well as another technique for improving the efficiency of heuristic search, action elimination.

4.5.1   ACTION ELIMINATION

The FIND-and-REVISE schema involves a certain implicit but computationally expensive step. It postulates that heuristic search algorithms should look for unconverged states in the current greedy graph $G_{s_0}^V$. The question that arises immediately is: how can we tell whether a given state s actually belongs to $G_{s_0}^V$? In general, this requires computing at least part of $G_{s_0}^V$, which, in turn calls for determining greedy actions in some fairly large set of states. In LAO*-like approaches, this set of states is the explicit graph. In the RTDP family, this set of states is the trajectory sampled in every iteration. Calculating greedy actions even in one state can be quite costly — it forces us to take a summation over all the successors of each action in that state.

Maintaining upper and lower bounds on each state value can help us alleviate the problem of finding greedy actions by letting us avoid evaluating many actions in a given state. To see how we can do it, suppose we maintain value functions V_l and V_u s.t. $V_l \leq V^*$ and $V^* \leq V_u$. We have already seen heuristic search algorithms that do this — BRTDP, FRTDP, and VPI-RTDP. Suppose that, during a visit to some state s we discover that $Q^{V_u}(s, a) < Q^{V_l}(s, a')$ for some pair of actions a, a'. This result would tell us immediately that

- Action a' currently cannot be the best action in s — even an optimistic estimate of its long-term cost ($Q^{V_l}(s, a')$) is higher than a pessimistic estimate for a ($Q^{V_u}(s, a)$).

- Moreover, action a' *cannot ever become* as good as a in s [22]!

The latter observation can be formalized as Theorem 4.15:

Theorem 4.15   Suppose that V_l and V_u are two value functions of an SSP MDP s.t. $V_l(s) \leq V^*(s)$ and $V_u(s) \geq V^*(s)$ for all states $s \in \mathcal{S}$. Furthermore, suppose that for a pair of actions $a, a' \in \mathcal{A}$ and some state $s \in \mathcal{S}$, $Q^{V_u}(s, a) < Q^{V_l}(s, a')$. Then a' is not recommended in s by any optimal policy π^*.

Thus, while running a FIND-and-REVISE algorithm with both an upper and a lower bound, whenever we detect that $Q^{V_u}(s, a) < Q^{V_l}(s, a')$ for two actions in a state s, we never have to evaluate a' in s again. It is sufficient to mark a' as *eliminated* in that state.

Heuristic search algorithms such as LAO* that, in their pure form, do not use upper bounds on the optimal value function can benefit from action elimination too if modified to use such a bound. The computational gains depend on the cost of computing two heuristics and maintaining the corresponding value functions. The ways of constructing upper bounds proposed in the literature [63;

140; 176], coupled with suitable admissible heuristics, typically yield large speedups across a range of domains. In fact, as we are about to see, even algorithms that do not fall under the heuristic search framework can exploit action elimination to their advantage.

4.5.2   FOCUSED TOPOLOGICAL VALUE ITERATION

Recall the Topological Value Iteration (TVI) algorithm from Section 3.6.1. It is a Partitioned Value Iteration algorithm that breaks up the MDP's transition graph into strongly connected components and runs VI on each of them in their topological order. Intuitively, TVI runs VI in the order of causal dependencies present in the MDP's state space. Its speedup over ordinary VI depends on its ability to separate the transition graph into pieces that are small. However, if the graph is densely connected, this may not be possible — the entire graph may be a single large component.

To do better than VI even on such MDPs, Focused Topological Value Iteration (FTVI) [63; 66] modifies TVI by employing action elimination tactics before computing the strongly connected components of the transition graph. In particular, FTVI starts with V_l and V_u bounds on the optimal value function and runs several LAO*-like heuristic search passes starting from the start state. These result in updates of V_l and V_u bounds in many states of the transition graph rooted at s_0.

Afterward, it eliminates provably suboptimal actions based on action elimination (Theorem 4.15). The elimination uses the informative bounds computed during the above heuristic search step. Finally, FTVI runs TVI on this pruned transition graph.

This approach also has a common-sense explanation — even though one state may be causally dependent upon another via *some* action, this may not be the case under the optimal policy. Action elimination removes the dependencies that do not actually matter, approximating the causality structure of the optimal policy. Doing so allows FTVI to outperform not only TVI but in some cases also dedicated heuristic search algorithms such as LRTDP.

4.6   COMPUTING ADMISSIBLE HEURISTICS

Generally, to provide good guidance, a heuristic needs to reflect knowledge about the problem at hand, and therefore needs to be chosen or derived in a domain-specific way. For instance, in a scenario where one needs to plan a safe cost-efficient route for an airplane, one might lower-bound the expected cost of getting to the goal by the cost of fuel the airplane would spend if it flew to the destination airport in a straight path with some tailwind.

Nonetheless, there are heuristics that attempt to derive state estimates purely from the specification of the MDP, without using any additional semantic information about it. While domain-specific heuristics may be more powerful, these *domain-independent* heuristics are a good option to fall back on when human intuition about the problem fails to produce a reasonable estimate for some state. Since domain-independent heuristics typically do not use semantic (i.e., human-interpretable) information about the problem, they often provide insights into the structure of the problem that humans would not be able to glean based on their intuition alone. To get the best of both worlds, practitioners can combine an arbitrary number of available domain-specific and domain-independent admissible heuristics by taking the maximum or some other admissible aggregate of their values for each state.

In this section, we discuss only admissible domain-independent heuristics. By Theorem 4.11, they guarantee asymptotic converge of heuristic search methods to an optimal policy. We point out, however, that inadmissible heuristics, while not providing optimality guarantees, are often more informative in practice, i.e., more capable of telling "good" states from "bad" ones. Thanks to this, inadmissible heuristics generally enable heuristic search algorithms to scale better, albeit at the price of solution optimality. We postpone the description of inadmissible heuristics until Section 6.3, where they are discussed as one of the techniques for solving MDPs efficiently but suboptimally.

4.6.1   ADAPTING CLASSICAL PLANNING HEURISTICS TO MDPS

Virtually all currently known domain-independent heuristics for MDPs work by taking an existing classical planning heuristic, relaxing the given MDP into a deterministic version called a *determinization* of the MDP, and applying the classical planning heuristic to the determinization. Determinizations can be constructed in a variety of ways, and we will see several of them when discussing approximate methods for solving MDPs in Chapter 6. However, both in approximation algorithms and in heuristics, the most widely used one is the *all-outcome determinization* [250].

Definition 4.16   All-Outcome Determinization. The all-outcome determinization of an SSP MDP $M = \langle \mathcal{S}, \mathcal{A}, \mathcal{T}, \mathcal{C}, \mathcal{G} \rangle$ is the SSP MDP $M_d^a = \langle \mathcal{S}, \mathcal{A}', \mathcal{T}', \mathcal{C}', \mathcal{G} \rangle$ s.t. for each triplet $s, s' \in \mathcal{S}$, $a \in \mathcal{A}$ for which $\mathcal{T}(s, a, s') > 0$, the set $\mathcal{A}'$ contains an action a' for which $\mathcal{T}'(s, a', s') = 1$ and $\mathcal{C}'(s, a', s') = \mathcal{C}(s, a, s')$.

Informally, M_d^a is just an MDP that has a separate deterministic action for each probabilistic outcome of every action of M. Thus, M_d^a is essentially a version of M where outcomes of M's actions can be controlled independently.

For a PPDDL-style factored MDP M, M_d^a is very easy to build. Recall from Section 2.5.2 that in a PPDDL-style factored representation, actions have the form $a = \langle prec, \langle p_1, add_1, del_1 \rangle, \cdots, \langle p_m, add_m, del_m \rangle \rangle$. Accordingly, M_d^a's action set consists of actions $a_i' = \langle prec, \langle 1.0, add_i, del_i \rangle \rangle$ for each i and each such a. The heuristics described below assume that the MDP is given in a PPDDL-style factored form.

4.6.2   THE h_{aodet} HEURISTIC

To get an idea of how M_d^a can help us estimate the value of a state s in M, consider the cheapest plan (a plan with the lowest cost) in M_d^a from s to the goal. Clearly, this plan is a possible way of getting from s to the goal in M as well, since it is a positive-probability sequence of M's action outcomes.

Therefore, the cost of this plan is a lower bound on the value of s in M — no policy in M can get the system from s to a goal at a smaller cost in expectation. Thus, we have the following theorem.

Theorem 4.17   The h_{aodet} heuristic, which estimates the value of s in M as the cost of the cheapest plan from s to the goal in the all-outcome determinization M_d^a of M, is admissible [229].

In some cases, h_{aodet} is a tight approximation to V^*; e.g., if $M_d^a = M$, h_{aodet} is the optimal value function!

To solve M_d^a optimally by finding a plan of the lowest cost from a given state to the goal, h_{aodet} can use any of the optimal classical algorithms, e.g., LRTA [139]. Unfortunately, for determinizations of factored MDPs this problem is PSPACE-complete [51]. In fact, the most efficient optimal deterministic planners, such as LRTA, themselves need heuristics. Due to this difficulty, most other MDP heuristics relax M_d^a into an even simpler problem, which makes them more efficient but occasionally less informative than h_{aodet}.

4.6.3   THE h_{max} HEURISTIC

Several well-known heuristics use the so-called *delete relaxation* of the all-outcome determinization. To keep the discussion focused, assume for now that the MDP we are trying to solve contains only positive literals in the goal conjunction and action preconditions (actions' effects may still contain negative literals though, i.e., have nonempty delete effects). The motivation for the delete relaxation is based on the observation that getting to the goal in a (deterministic) factored problem can be viewed as achieving one-by-one the set of literals $G_1, \ldots, G_n$ that form the goal conjunction. This is difficult because once one has achieved one of these literals, say, G_i, to achieve another, G_j, one may first have to "undo" G_i, achieve G_j, and then re-achieve G_i again. Thus, in a sense, the agent's progress toward the goal is not monotonic.

The ultimate culprit is the actions' delete effects, discussed in Section 2.5. Because of them, an action may achieve one "desirable" literal and simultaneously remove another. Thus, if MDPs did not have delete effects, finding *a* goal plan in them would be significantly easier.

The delete relaxation turns the deterministic problem at hand (which, in our case, is a determinization of an MDP) into such a "delete-free" problem by taking each action $a = \langle prec, \langle 1.0, add, delete \rangle \rangle$ of the determinization and replacing it with $a' = \langle prec, \langle 1.0, add, \bigwedge \emptyset \rangle \rangle$. We denote the set of actions of the delete relaxation as $\mathcal{A}^r$. The optimal solution to this problem has an important property.

Theorem 4.18   The cost of an optimal goal plan from state s in the delete relaxation of a given deterministic planning problem is a lower bound on the cost of an optimal goal plan from s in the deterministic problem itself [34].

Via Theorem 4.17, this result implies that the heuristic that estimates the value of a state s in M as the cost of the optimal plan from s to the goal in the delete relaxation of the determinization is admissible for M as well.

However, as it turns out, finding an *optimal* plan (or its cost) in the delete relaxation is NP-hard [30] — the delete relaxation is not much easier to solve than the determinization itself! However, we can bound the cost $C(s, \mathcal{G})$ of the optimal plan from s to the goal in the delete relaxation from below. To do so, let us assume that the cost of an action in the original MDP M does not depend on the state in which the action is used, and thus can be captured by a single value $Cost(a)$. Then we can bound $C(s, \mathcal{G})$ with $C(s, \mathcal{G}) \geq \max_{1 \leq i \leq n} C(s, G_i)$, where $G_1, \ldots G_n$ are the literals in the goal conjunction and $C(s, G_i)$ is an estimated cost of a plan that achieves only literal G_i. Therefore, we define

$$h_{max}(s) = \max_{1 \leq i \leq n} C(s, G_i), \tag{4.1}$$

where $C(s, G_i)$ can be approximated as

$$C(s, G_i) = \begin{cases} 0 & \text{if } G_i \text{ holds in } s, \\ \min_{a \in \mathcal{A}^r \text{ s.t. } G_i \text{ is in } add(a)} Cost(a) + C(s, prec(a)) & \text{if } G_i \text{ does not hold in } s, \end{cases} \tag{4.2}$$

where

$$C(s, prec(a)) = \max_{L_i \text{ s.t. } L_i \text{ holds in } prec(a)} C(s, L_i) \tag{4.3}$$

The heuristic that estimates the value of a state s in M as $\max_{1 \leq i \leq n} C(s, G_n)$ is called h_{max} [30] and is admissible, since the estimate $C(s, G_i)$ of achieving any single goal literal cannot be higher than achieving all of the goal literals. Applying these equations recursively is similar to a shortest-path algorithm for graphs.

Returning to the assumption of purely non-negative literals in the goal and action preconditions, h_{max} can still be used even on problems where this assumption does not hold. One way to apply h_{max} to such problems is to simply drop the negative literals from the preconditions and the goal. A more principled way is to compile the problem into the STRIPS formulation [91], which has only positive literals in the goal and action preconditions. This is done by replacing each instance of each negative literal $\neg L$ in the goal and preconditions of actions in the determinization by a newly introduced positive literal L', adding L' to the add effect of every action whose delete effect contains $\neg L$, and removing $\neg L$ from all the delete effects.

The strength of h_{max} lies in its ease of computation. However, sometimes it is not very informative, since $\max_{1 \leq i \leq n} C(s, G_n)$ is often a rather loose bound on the cost of the optimal solution of M_d^a.

4.7   HEURISTIC SEARCH AND DEAD ENDS

In Section 3.10 we explored how we can extend the SSP MDP definition to allow goal-oriented MDPs to have dead ends, a common type of states in many scenarios. The new fSSPDE MDP class we defined dispensed with SSP MDP's requirement of a complete proper policy existence. Instead, it puts a user-defined cap on the expected cost of reaching the goal from any state. This ensures that all policies in fSSPDE MDPs have finite expected costs and lets the existing SSP MDP algorithms solve fSSPDE problems.

Do heuristic search algorithms also apply to fSSPDE MDPs if the latter are equipped with an initial state? The short answer is "yes." However, as we are about to see, the knowledge of the initial state sometimes allows us to solve goal-oriented MDPs without assuming the existence of a complete proper policy *and* without artificially limiting the cost of every policy the way the fSSPDE definition does. Thus, the presence of an initial state lets us relax the requirements of both the SSP and fSSPDE classes.

4.7.1   THE CASE OF AVOIDABLE DEAD ENDS

To see how we can do this, observe that while an optimal solution to an SSP MDP is a complete policy, an optimal solution to an SSP_{s_0} MDP is a partial one. Therefore, if an SSP_{s_0} MDP has a proper policy *rooted at* s_0 and every improper policy incurs an infinite expected cost, an optimal solution is guaranteed to exist without any additional conditions on dead ends. Intuitively, this is true because even if dead ends are present, at least one policy for the MDP, the proper one, avoids them with 100% probability. The class of SSP MDPs with avoidable dead ends formalizes this idea:

Definition 4.19   Stochastic Shortest-Path MDP with Avoidable Dead Ends. A *stochastic shortest-path path MDP with avoidable dead ends (SSPADE)* is a tuple $\langle \mathcal{S}, \mathcal{A}, \mathcal{T}, \mathcal{C}, \mathcal{G}, s_0 \rangle$, where $\mathcal{S}, \mathcal{A}, \mathcal{T}, \mathcal{C}, \mathcal{G}$, and s_0 are as in the definition of an SSP_{s_0} MDP (Definition 4.3), under two conditions:

- There exists at least one policy closed w.r.t. s_0 that is proper at s_0,

- For every improper stationary deterministic Markovian policy π, for every $s \in \mathcal{S}$ where π is improper, $V^{\pi}(s) = \infty$.

To emphasize it once again, the SSPADE MDP definition has only one distinction from that of SSP_{s_0} MDPs. The former requires that a proper policy exist only w.r.t. the initial state, not everywhere. As a result, SSPADE MDP may have dead ends as long as they are avoidable from s_0.

The existence of an optimal solution does not imply that it is easy to find, *per se*. Nonetheless, in the case of SSPADE MDPs we can prove a result similar to Theorem 4.11.

Theorem 4.20   For an SSPADE MDP, if FIND-and-REVISE is initialized with an admissible heuristic, terminated when all the states in the value function's greedy graph rooted at s_0 are ϵ-consistent, and its FIND procedure is systematic, the value function computed by FIND-and-REVISE approaches the optimal value function over all states reachable from s_0 by an optimal policy as ϵ goes to 0 [136].

Unfortunately, this does not mean that all the heuristic search algorithms developed for SSP_{s_0} MDPs, such as LRTDP or LAO*, will eventually arrive at an optimal policy on any SSPADE problem. The subtle reason for this is that the FIND procedure of some of these algorithms is *not* systematic when dead ends are present. Let us consider the applicability to SSPADE MDPs of the most widely used heuristic search algorithms, LAO*, ILAO*, RTDP, and LRTDP:

- **LAO***. Suppose that the state chosen for expansion on line 10 of LAO*'s pseudocode (Algorithm 4.2) is a dead-end state s_d whose only a transition is a deterministic one back to itself. This triggers the execution of PI on line 17 on a set including s_d. PI fails to halt on dead ends such as s_d, because the value of any policy on them is infinite and hence is impossible to arrive at in a finite amount of time. To "fix" LAO* for SSPADE MDPs, one can, for instance, artificially limit the maximum number of iterations in PI's policy evaluation step.

- **ILAO***. Unlike LAO*, this algorithm converges on SSPADE problems without any modification. It runs PI on the states in the current greedy graph until either PI converges *or the greedy graph changes*. Now, the greedy graph always includes the initial state s_0. By the SSPADE MDP definition, there is always a policy that has a finite value at s_0, and any policy that can lead to a dead end from s_0 has an infinite value at s_0. Therefore, running PI on a greedy graph that includes a dead end must cause an action change at s_0 (or some other state) by making the policy that visits a dead end look bad.

- **RTDP**. Observe that if an RTDP trial enters a dead end whose only transition leads deterministically back to that state, the trial will not be able to escape that state and will continue forever. Thus, RTDP may never approach the optimal value function no matter how much time passes. Artificially limiting the trial length (the number of state transitions in a trial) with a finite N works, but the magnitude of N is very important. Unless $N \geq |\mathcal{S}|$, convergence of RTDP to the optimal value functions cannot be *a priori* guaranteed, because some states may not be reachable from s_0 in less than N steps and hence will never be chosen for updates. On the other hand, in MDPs with large state spaces setting $N = |\mathcal{S}|$ may make trials wastefully long.

- **LRTDP.** LRTDP fails to converge to the optimal solution on SSPADE MDPs for the same reason as RTDP, and can be amended in the same way.

As illustrated by the examples of the above algorithms, SSPADE MDPs do lend themselves to heuristic search, but designing FIND-and-REVISE algorithms for them warrants more care than for SSP_{s_0} MDPs.

4.7.2   THE CASE OF UNAVOIDABLE DEAD ENDS

Although SSPADE MDPs can be solved with largely the same heuristic search algorithms as presented earlier in this chapter, the adherence of an MDP to the SSPADE definition is often hard to verify. This applies to SSP MDPs as well — indeed, in both cases the fact that a proper policy exists needs to be somehow inferred from domain-specific knowledge *before* the MDP is actually solved. Because of this, in practice it may be more convenient to assume that an MDP may have dead ends that are *unavoidable* from s_0 and use algorithms capable of handling these problems.

Notice that once we admit the possibility of unavoidable dead ends in goal-oriented MDPs with an initial state, we run into the same difficulties as already discussed in Section 3.9 for SSP MDPs with *no* initial state. All policies in MDPs with unavoidable dead ends have an infinite expected cost at s_0 if evaluated under the expected linear additive utility criterion, making the comparison of policies by their values meaningless. In Section 3.9 we outlined a general way of altering the evaluation criterion in order to render policies comparable in a meaningful way — imposing a penalty $\mathcal{P}$ on visiting a dead end state. Using this idea, we formulate an analog of Definition 3.26 for goal-oriented MDPs with an initial state:

Definition 4.21   Stochastic Shortest-Path MDP with Unavoidable Dead Ends. A *stochastic shortest path MDP with unavoidable dead-ends (SSPUDE)* is a tuple $\langle \mathcal{S}, \mathcal{A}, \mathcal{T}, \mathcal{C}, \mathcal{G}, \mathcal{P}, s_0 \rangle$, where $\mathcal{S}$, $\mathcal{A}, \mathcal{T}, \mathcal{C}, \mathcal{G}$, and s_0 are as in the definition of an SSP_{s_0} MDP (Definition 4.3), and $\mathcal{P} \in \mathbb{R}^+ \cup \{+\infty\}$ denotes the penalty incurred when an agent decides to abort the process in a non-goal state, under one condition:

- For every improper stationary deterministic Markovian policy π, for every $s \in \mathcal{S}$ where π is improper, the value of π at s under the expected linear additive utility is infinite.

If $\mathcal{P} < \infty$, the MDP is called a *finite-penalty SSPUDE (fSSPUDE)* MDP. If $\mathcal{P} = \infty$ it is called an *infinite-penalty SSPUDE (iSSPUDE)*.

For finite-penalty scenarios, we have a result similar to Theorem 3.27.

Theorem 4.22 $SSP_{s_0} = fSSPUDE$.

By this equivalence we mean that every fSSPUDE MDP can be converted to an SSP_{s_0} MDP with the same set of optimal policies as the original fSSPUDE instance, and vice versa.

An immediate corollary of this theorem is the fact that all FIND-and-REVISE algorithms for SSP_{s_0} MDPs work for fSSPUDE MDPs without a need for any changes.

CHAPTER 5

Symbolic Algorithms

In this chapter we discuss those MDP algorithms that use compact value function representations, in particular, decision diagrams, such as *binary decision diagrams (BDD)* and *algebraic decision diagrams (ADD)* [9; 44]. These representations (and associated algorithms) are often referred to as *symbolic* or *structured* representations (algorithms). Symbolic techniques complement the algorithms from previous chapters, and several previous algorithms have (or can have) symbolic counterparts for added efficiency.

The key idea of symbolic algorithms is to compactly store the MDP tables (value functions, transition functions, costs, etc.) using decision diagrams. These representations naturally cluster those states that behave similarly, e.g., have the same values. If a subset of states behaves similarly, then instead of redundant backups for each state, it is sufficient to consider the whole subset at once and update all values by just a single backup. This speeds up all the VI-based algorithms. Moreover, the compact representation saves valuable space. Thus, symbolic algorithms often scale far better than their *flat* counterparts (i.e., where the functions are stored as tables).

The main requirement for applying symbolic techniques is that we have access to a factored MDP (see Definition 2.22), i.e., the state space $\mathcal{S}$ is factored into a set of d state variables $\mathcal{X} = \{X_1, X_2, \ldots, X_d\}$. Moreover, as in the previous chapter, we assume that all variables X_i are Boolean. This is not a severe restriction, since a variable with finite domains could be converted into multiple Boolean variables.

It is important to note that a symbolic representation, such as using ADDs, is an instance of a factored representation. A factored representation exploits the knowledge of the factored MDP to represent a function (often, more compactly). Other factored representations have also been studied within MDPs, for instance the sum of basis function representation. We discuss that in Section 6.4.2.

We first describe the ADD representation and introduce ADD manipulation algorithms. We then focus on the two most popular algorithms that use the ADD representation for solving MDPs – SPUDD and Symbolic LAO*. In the end, we briefly discuss a few other algorithms that use ADD representations.

5.1 ALGEBRAIC DECISION DIAGRAMS

ADDs offer a compact representation for functions of the form $\mathcal{B}^d \to \mathbb{R}$ (where $\mathcal{B}$ is the set $\{0, 1\}$). BDDs represent functions of the form $\mathcal{B}^d \to \mathcal{B}$. For all practical purposes, a BDD is a special case of an ADD – so we focus on the latter.

Definition 5.1 Algebraic Decision Diagram. An *Algebraic Decision Diagram* is defined as a directed acyclic graph, with two types of nodes – internal nodes and leaves. Each internal node v has a label $var(v) \in \mathcal{X}$ and has exactly two outgoing edges labeled $hi(v)$ and $lo(v)$. A leaf node or a terminal node is labeled with $val(v) \in \mathbb{R}$. An ADD has exactly one *root*, i.e., a node with no incoming edges.

Figure 5.1 illustrates an example ADD, which represents the function f shown on the left in the figure. This ADD has four internal nodes and three leaves. The hi and lo edges from an internal node are shown as solid and dashed lines, and represent the variable's true and false values respectively. The leaves of an ADD (square nodes in the figure) store real numbers and represent the function values.

The semantics of an ADD is straightforward – given an assignment to all variables in $\mathcal{X}$, start from the root node and take appropriate edges forward based on X_i's given values. The resulting leaf node represents the function's value. For example, for the ADD in Figure 5.1, $f(0, 0, 1)$ can be computed by successively taking the dashed, dashed and solid edges starting at the root, X_1, to arrive at the value of 2.

The key benefit of an ADD representation is that it can be much more compact than the original table representation. It is effective in cases where the value function has some underlying structure, for instance, some variables may be important only in a part of the state space (also known as *context-specific independence*). For example, in Figure 5.1, if X_1 and X_2 are both true, then f's value is independent of X_3. Here, the ADD does not enumerate the two similar cases, and, instead, aggregates both of them into a single value bin. Notice that the same does not hold for a flat representation, which enumerates all of the exponential number of possible assignments.

X_1	X_2	X_3	$f(X)$
0	0	0	0
0	0	1	2
0	1	0	2
0	1	1	2
1	0	0	0
1	0	1	2
1	1	0	5
1	1	1	5

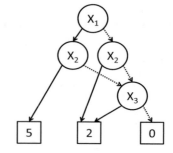

Figure 5.1: Example Algebraic Decision Diagram. The table shows the original function in a flat representation and the graph is its resulting reduced ADD representation.

The other important characteristic of ADDs is the existence of efficient algorithms for function manipulation and composition. For example, combining two functions (adding, multiplying, max etc.) can be performed directly within the ADD graph structures. The efficiency of ADD algorithms hinges on a predefined variable order. A specific order for the state variables is chosen in advance, and a state variable is never repeated in a path. Such ADDs are called *ordered algebraic decision diagrams*.

We first describe two important ADD algorithms 'REDUCE' and 'APPLY' in detail.

5.1.1   THE REDUCE OPERATOR

The goal of the 'REDUCE' algorithm is to make an ordered ADD more compact without affecting its function evaluation. It basically looks for redundancies in the data structure and removes them. It checks for three sources of redundancy:

1. **Duplicate Leaves:** If many leaves have the same value, then they are duplicates, and all except one are eliminated. All incoming edges are redirected to the remaining leaf.

2. **Duplicate Internal Nodes:** Two internal nodes v_1 and v_2 are duplicates, if they have the same variable and children, i.e., $var(v_1) = var(v_2)$, $hi(v_1) = hi(v_2)$, and $lo(v_1) = lo(v_2)$. 'REDUCE' eliminates one of the nodes and redirects all incoming edges to the other one.

3. **Redundant Internal Node:** An internal node v is redundant if both its children are the same, i.e., $hi(v) = lo(v)$. In this case, v is eliminated and its incoming edges are redirected to its child.

The 'REDUCE' algorithm eliminates these redundant nodes in a bottom-up fashion, making this operation efficient. Figure 5.2 illustrates the steps in reducing the full-tree representation of the function from Figure 5.1.

We can prove that after the 'REDUCE' operator, an ordered ADD has a *canonical* representation, i.e., two ADDs encoding the same function will be isomorphic, under a common variable order. This makes checking for function equivalence straightforward. Such ADDs are called *reduced ordered algebraic decision diagrams*. These are the ADDs of choice for implementing MDP algorithms.

5.1.2   THE APPLY OPERATOR

The 'APPLY' operator is the basic function composition algorithm over ordered ADDs. It takes two ADDs $\overline{F_1}$ and $\overline{F_2}$, which represent functions F_1 and F_2, and a composition operator op to compute $F_1 \ op \ F_2$. The operator op may be sum, product, max, etc. For a BDD it may also be a Boolean operation like XOR, AND, etc. Note that we use a bar to denote the ADD representation of a function ($\overline{F_1}$ denotes ADD of function F_1).

The following set of recursive equations are used to implement the 'APPLY' algorithm over two ordered ADDs that have the same variable order. Here MK(var, hi, lo) represents construction of a node in the ADD with variable var and hi and lo as its two children nodes.

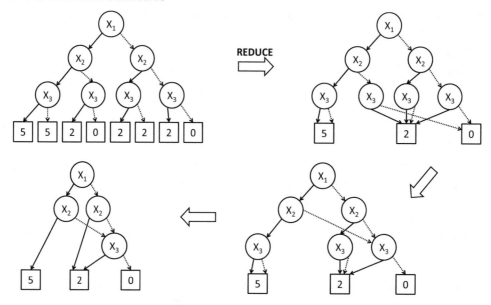

Figure 5.2: Application of 'REDUCE' operator on the example of Figure 5.1. First, it eliminates duplicate leaves. Then at each depth, first it eliminates duplicate internal nodes and then redundant nodes.

$$\text{APPLY}(\overline{F_1} \; op \; \overline{F_2}) = \text{REDUCE}(\text{APP}(\overline{F_1}.root \; op \; \overline{F_2}.root))$$

$$
\begin{array}{llll}
\text{APP}(v_1 \; op \; v_2) & = & val(v_1) \; op \; val(v_2) & \text{if } v_1 \text{ and } v_2 \text{ are leaves} \\
& = & \text{MK}(var(v_1), \text{APP}(hi(v_1) \; op \; hi(v_2)), \text{APP}(lo(v_1) \; op \; lo(v_2))) & \text{if } var(v_1) = var(v_2) \\
& = & \text{MK}(var(v_1), \text{APP}(hi(v_1) \; op \; v_2), \text{APP}(lo(v_1) \; op \; v_2)) & \text{if } var(v_1) < var(v_2) \\
& = & \text{MK}(var(v_2), \text{APP}(v_1 \; op \; hi(v_2)), \text{APP}(v_1 \; op \; lo(v_2))) & \text{if } var(v_1) > var(v_2)
\end{array}
$$

'APPLY' follows two main steps. It first applies the operator *op* by traversing the two ADDs simultaneously and creating a new value for each sub-case. Next, it reduces the resulting ADD so that it is as compact as possible.

Example 5.3 illustrates the end result of adding two functions. Notice that the resulting ADD is smaller than one of the original functions, due to an application of the 'REDUCE' step on an intermediate larger ADD. Applying an operator between two functions takes worst-case $O(|\overline{F_1}||\overline{F_2}|)$ time, where $|\overline{F_1}|$ is the number of nodes in the ADD representation of F_1.

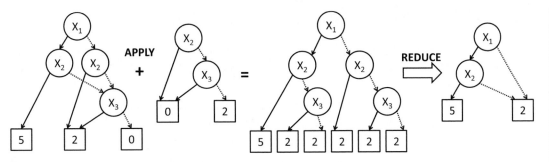

Figure 5.3: The result of using the 'APPLY' operator to add two functions.

5.1.3   OTHER ADD OPERATORS

An operation that is easy to perform with ADDs is 'RESTRICT' – restricting the function representation by choosing a particular value for a variable. It is performed by removing each mention of that variable and replacing it with one of the children in the ADD (depending on the value with which we want to restrict the variable).

Another operation we will need is marginalizing out a variable (let us call it the 'MARGINAL-IZE' operator). This is achieved by removing the node and replacing it by the sum of its two children, which can be easily computed by using the 'APPLY' operator.

Last but not the least, it is also easy to convert a flat representation of a function into an ADD: choose a variable order, construct a full decision tree enumerating all possible states and then apply the 'REDUCE' operator to reduce the size of the ADD. An example is illustrated in Figure 5.2. To study the ADD (BDD) algorithms in more detail, please refer to the original papers [9; 45].

It is important to note that the variable order plays an enormous role in determining the size of an ADD. The same function may take linear space using one variable order and exponential using another. Finding the optimal variable-order is NP-hard [28]. In practice, heuristic approaches (local search, simulated annealing) are used to choose a good variable order.

A binary decision diagram (BDD) is typically used to represent logic functions. BDD operators are commonly logic operators like AND, OR, XOR, etc. A BDD can also be used to represent a set of states (using an indicator function), in which case they also allow set-theoretic operators like intersection, union, and set difference. They also allow other complex operations such as existential quantification. We revisit this in Section 5.3.

Finally, one of the most valuable aspects of ADDs and BDDs is that an efficient implementation of these operations is available in the CUDD package [220]. Most of the popular algorithms have been tested using CUDD.

5.2   SPUDD: VALUE ITERATION USING ADDS

The use of symbolic representations helps optimize many iterative algorithms we covered in previous chapters. We first discuss SPUDD [111], which is an implementation of Value Iteration (VI) using ADDs. The main idea is to perform each iteration of VI with reduced ordered ADDs. This optimizes the sweep over the states, since similarly behaving states, i.e., ones that may have their values updated similarly in Bellman backups, need not be considered separately.

Why would multiple states behave in a similar fashion? Often, several state variables play little or no role in determining a state's value function. For instance, if a Mars rover's goal is to do an experiment on a particular rock then the location of other rocks is irrelevant. So, the states that agree on all variables except the locations of irrelevant rocks will have the same values. Similarly, if a taxi's objective is to drop off passengers at their desired location, and the taxi has already picked up the passengers, then their original wait locations are no longer relevant in determining the state values. If such similar states have equal current values, then they will get updated to same next values after each VI iteration. An ADD representation will cluster all these states together. When making a sweep over the state space, it will consider the whole cluster once, instead of going over each of these states separately. This will save both memory and time.

The first step is to represent the various tables used in VI using an ADD. We need to represent the following tables: the current and next value functions $V_n(s)$ and $V_{n+1}(s)$, the transition function $\mathcal{T}(s, a, s')$ and the action costs $\mathcal{C}(s, a, s')$.

Assuming a state to be composed of d Boolean state variables, $V_n(s)$ is just a function $\mathcal{B}^d \to \mathbb{R}$. This can be simply represented as an ADD of depth d. We use a bar to suggest an ADD representation, e.g., $\overline{V}_n$ denotes the ADD for V_n.

Representing the transition function is a bit trickier. It needs to represent a function $\mathcal{B}^d \times \mathcal{A} \times \mathcal{B}^d \to [0, 1]$. Since the set of actions is not a Boolean this is most easily represented by $|\mathcal{A}|$ ADDs, each representing a transition for each action. We use $\overline{\mathcal{T}}^a$ to denote the transition function for action a, which is therefore a function $\mathcal{B}^{2d} \to \mathbb{R}$. The internal nodes are $X_1, \ldots, X_d, X'_1, \ldots, X'_d$ (primed variables represent variables in the state s'). The cost functions are represented analogously to the transitions.

Once we have ADDs for $\mathcal{T}, \mathcal{C}$, and V_0 we are ready to execute VI using ADDs. Algorithm 5.1 describes the pseudo-code. We start with $\overline{V}_0$, the ADD representation of V_0. We then execute the VI iterations, except instead of going over all states explicitly, we perform the computations for all states at once. For each action we need to compute $Q_{n+1}(s, a) = \sum_{s' \in \mathcal{S}} \mathcal{T}(s, a, s')[\mathcal{C}(s, a, s') + V_n(s')]$.

To calculate the right hand side, we first add $\mathcal{C}(s, a, s')$ and $V_n(s')$. Both $\mathcal{C}$ and V_n are represented as ADDs, so this is just an 'APPLY' operation. Since we need $V_n(s')$, we rename all variables in $\overline{V}_n$ by their primed counterparts. We call this intermediate diagram $\overline{dd1}$ (line 7). Next, we multiply $\overline{dd1}$ with the transition function $\overline{\mathcal{T}}^a$. This is also an 'APPLY' operator. We now need to sum over all s'. We obtain this by using 'MARGINALIZE' over all primed variables. The last step in the loop requires computing $\overline{V}_{n+1}$ as a min over the ADDs for $Q_{n+1}(s, a)$. In essence, the whole set of Bellman backups in an iteration is compressed in a few ADD operations.

To compute the residual we need to track the max |leaf value| in $\overline{V}_{n+1} - \overline{V}_n$. To compute the final greedy policy we just repeat one more iteration and this time while computing the min (line 14), also annotate which action the value came from. This results in an ADD representation of the returned policy.

Value Iteration implemented with ADDs has similar convergence, termination, optimality, and monotonicity results as the original VI (Section 3.4). This is because we have not changed the dynamic programming computations, just clustered the appropriate states together, so that some redundant computations could be avoided.

Algorithm 5.1: SPUDD: Value Iteration with ADDs

1   initialize $\overline{V}_0$
2   $n \leftarrow 0$
3   **repeat**
4      $n \leftarrow n + 1$
5      $\overline{V'}_n \leftarrow \overline{V}_n$ with all X_is in renamed as X'_i
6   **foreach** $a \in \mathcal{A}$ **do**
7          $\overline{dd1} \leftarrow \text{APPLY}(\overline{C}^a + \overline{V'}_n)$
8          $\overline{dd2} \leftarrow \text{APPLY}(\overline{dd1} \times \overline{\mathcal{T}}^a)$
9   **foreach** X'_i **do**
10             $\overline{dd2} \leftarrow \text{MARGINALIZE}(\overline{dd2}, X'_i)$
11   **end**
12          $\overline{Q}^a \leftarrow \overline{dd2}$
13   **end**
14      $\overline{V}_{n+1} = \min_{a \in \mathcal{A}}\{\overline{Q}^a\}$ using successive 'APPLY' operators
15 **until** $\max_{s \in \mathcal{S}} |\overline{V}_{n+1} - \overline{V}_n| < \epsilon$;
16 compute greedy policy

To use VI with ADDs we need to be careful about a potential problem. The ADDs, especially the transition functions (and costs, if they depend on next states), have $2d$ depth and can exhaust the memory. Researchers have investigated this issue in specific cases.

In many MDPs, actions costs do not depend on the next state, and depend only on the state where an action is applied. This makes the cost-function ADD much smaller. Moreover, often, the transition function of the factored MDP can be compactly described using a dynamic Bayes net (DBN) representation [71]. In a DBN, the transition of each variable is expressed independently (see Section 2.5.3). The easiest case is when each variable X'_i depends only on the previous state $\mathcal{X}$ and the action, but does not depend on other X'_js. In other words, in this case, the actions have no correlated effects.[1] In addition, let us assume that each variable depends only on a few variables, instead of the entire previous state. In such a case, we can construct d smaller ADDs, one for the transition of each variable. We can multiply each ADD with $\overline{dd1}$ and sum out that primed variable

---

[1] DBNs do allow correlated effects by introducing dependencies between two primed variables. We do not discuss those for ease of explanation.

right then and there. This successive computation is an alternative method to compute $\overline{Q}^a$ (see line 12), but no ADD grows too large.

Notice that this memory saving comes at the cost of some repetitive computation (since the transition probability ADDs for each variable are repeatedly multiplied). This suggests a tuning knob, where the variables X_i's are divided into partitions such that the ADDs grow large enough to fit but not exceed the available memory. This achieves the best time-space tradeoff [111].

5.3   SYMBOLIC LAO*

Symbolic LAO*, as the name suggests, is a symbolic extension of the LAO* algorithm [85]. Readers may wish to remind themselves of the LAO* algorithm from Section 4.3.1, which is a heuristic search algorithm that uses a heuristic h to search only a subset of the state space, starting from the initial state s_0.

Symbolic LAO* combines the power of heuristic search (no time wasted on states unreachable from the start state) with that of ADD representations (efficient memory use, computations saved on similarly behaving states).

At first glance, it is not obvious how to combine the two. ADDs by nature represent a function for the whole state space and perform computations over the whole of it at once. On the other hand, heuristic search specifically aims at avoiding computations for the entire state space. The combination of ADDs and heuristic search is achieved by letting heuristic search decide on the subset of states to explore and letting ADDs optimize the computations for *that* subset of the state space. We first discuss some ADD/BDD operations necessary for Symbolic LAO*.

Masked ADD: To enable ADD computation over only a subset of the state space we need to define a way to avoid spending time in the rest of the state space. This is achieved by the notion of *masking an ADD*. Given an ADD $\overline{A}$ and a subset of states S we compute $mask_S(\overline{A})$. This is a function with the same values as $\overline{A}$, except for the states not in S, for which the value is mapped to zero. To compute this, we define an indicator BDD which has a value 1 for each state in S and zero otherwise. We abuse the notation a little and use $\overline{S}$ to denote this indicator BDD. Multiplying the two ADDs results in the masked ADD, i.e., $mask_S(\overline{A}) = \text{APPLY}(\overline{S} \times \overline{A})$. It is easy to see that such a masked ADD might have dramatically fewer nodes, especially when S is small.

The Image BDD: Another useful BDD is the state-transition BDD, $\overline{Tr}^a$. Recall that, in the previous section, $\overline{T}^a$ represented the ADD to store the complete transition function for action a, with probabilities of transitions as leaves. $\overline{Tr}^a$ is a simplification where the leaves are only 0 or 1. A value of 1 represents that the transition between the two states is possible using a, with a non-zero probability.

The state-transition BDD helps in computing the *image* of a set of states S through action a, denoted by $Image^a(\overline{S})$. The image of a set of states S computes all the states that can be reached in one step by a starting from a state in S. The image is computed as an indicator BDD, where a leaf of 1 represents that such a state can be reached in one step from S.

This computation proceeds as follows. The BDD $mask_S(\overline{Tr}^a)$ has $2d$ variables (has both Xs and X's) and represents the one-step domain transitions using a, but only when the predecessor state is in S. For $Image^a$, we need the set of next states. We apply a 'MARGINALIZE'-like operation over X_is on this BDD, except that the leaf values are orred instead of summed. In other words, it removes the unprimed X variables and only those combinations of X' variables are left with value 1 that originally had a value 1 in conjunction with some X variables. Therefore, this BDD represents all the next states reachable from S. Since we want the final output in terms of unprimed variables, we just rename all X's to Xs and obtain $Image^a(\overline{S})$.

Partial Policy BDD: Finally, we need to keep track of the current greedy partial policy. We define $|\mathcal{A}|$ indicator BDDs, $\overline{S_a}$ where S_a represents the set of states in which a is the current best action. Having defined these concepts, we are now ready to implement LAO* using ADDs.

The Symbolic LAO* Algorithm: The pseudo-code for Symbolic LAO* is presented in Algorithm 5.2. Recall that LAO* operates in two steps. First it finds the fringe states for the current greedy partial policy and expands one or more of those. Then it performs VI to update the value function on a relevant subset of states.

Symbolic LAO* follows the same principles, except that in each expansion it updates *all* the fringe states, instead of just one. In that sense, it is closer to ILAO*. The VI step is computed only over the states in the greedy policy graph. This is achieved by masking each ADD involved in the computation.

Lines 5 to 16 in pseudo-code describe the method to find fringe states in the greedy policy graph ($\overline{F}_{gr}$). It successively gathers all next states in the greedy policy starting at s_0. This computes the interior states in the current greedy policy graph,($\overline{I}_{gr}$). To operationalize this, it uses the $Image$ operator that we described above (line 10). In computing this it uses the masking operator to ensure that the greedy policy is computed only for the subset of states already expanded (line 13). The new fringe states are added to the to $\overline{I}_{gr}$. It then needs to compute the new values for this greedy policy graph. This is computed using VI over ADDs, i.e., the SPUDD algorithm from the previous section.

Similar to original ILAO*, instead of running a full VI until convergence, one or a smaller number of iterations are preferred. This results in a symbolic version of ILAO*, which gives the best overall performance. Symbolic ILAO* was the top-performing optimal planner and achieved the second place in the 1st International Probabilistic Planning Competition in 2004 [254].

5.4 OTHER SYMBOLIC ALGORITHMS

ADD-based representations complement the enhancements discussed in the past two chapters. Thus, almost all dynamic programming-based MDP algorithms can be implemented using symbolic approaches. For instance, researchers have investigated structured policy iteration [38; 39]. This algorithm predates SPUDD and does not use ADDs; instead, it uses a precursor tree-based representation for the value functions. These tree representations are not as compact as ADDs, but the algorithms can be easily extended to ADDs in a straightforward way.

Algorithm 5.2: Symbolic LAO*

1 $\overline{V_I} \leftarrow h$ //initialize value function based on the heuristic
2 $\overline{S} \leftarrow \emptyset$ //initialize the set of states in the explicit graph
3 $\forall a \overline{S_a} \leftarrow \emptyset$ //initialize the set of states in greedy policy graph with greedy action a
4 **repeat**
5 | $\overline{I}_{gr} \leftarrow \emptyset$ //initialize interior states in the greedy policy graph
6 | $\overline{F}_{gr} \leftarrow \emptyset$ //initialize the set of fringe states in the greedy policy graph
7 | $\overline{from} \leftarrow \{s_0\}$
8 | //construct the sets of interior and fringe states in the greedy policy graph starting at s_0
9 | **repeat**
10 | | $\overline{to} \leftarrow \bigcup_{a \in \mathcal{A}} Image^a(mask_{S_a}(\overline{from}))$
11 | | $\overline{F}_{gr} \leftarrow \overline{F}_{gr} \cup (\overline{to} \setminus \overline{S})$
12 | | $\overline{I}_{gr} \leftarrow \overline{I}_{gr} \cup \overline{from}$
13 | | $\overline{from} \leftarrow mask_S(\overline{to}) \setminus \overline{I}_{gr}$
14 | **until** $\overline{from} = \emptyset$;
15 | $\overline{E}_{gr} \leftarrow \overline{I}_{gr} \cup \overline{F}_{gr}$ //union of interior and fringe states in greedy policy graph
16 | $\overline{S} \leftarrow \overline{S} \cup \overline{F}_{gr}$ //update the explicit graph by adding the new fringe states
17 | run Value Iteration (Algo 5.1) masked with states in E_{gr}
18 | extract greedy policy and update all $\overline{S_a}$
19 **until** $\overline{Fgr} = \emptyset$ and small residual;
20 return greedy policy

Similarly, other heuristic search algorithms have also been extended to symbolic implementations. A prominent example is Symbolic RTDP [86]. RTDP is by nature an online algorithm simulating one action at a time. The symbolic version of RTDP generalizes the current state s into a set of states that share structural (or value-based) similarity with it. Using the masking operator, a Bellman backup is performed on this set of states. Later, as with RTDP, the greedy action for s is simulated and a new current state reached.

In essence, Symbolic RTDP is able to generalize the information from the current Bellman backup onto several similar states. Of course, the key question is how to generalize the current state to a set? Two approaches have been proposed. The first is to generalize a state to all states that have the same (or similar) current value. This allows the joint update of all states whose values may remain similar throughout the algorithms. The other approach exploits the states' structural similarity. A backup at s backs up values of all successor states s'. This approach collects all predecessors of s', thus backing values of s' to all states that can use it. In the original experiments, value-based generalization performed relatively better between the two, but Symbolic LAO* still outperformed Symbolic RTDP. Recent work has extended Bounded RTDP to Symbolic BRTDP, thus allowing the use of both lower and upper bounds with symbolic approaches [76].

Researchers have also proposed structured versions of Generalized Prioritized Sweeping (see Section 3.5.1), in which the priority queue is a queue of abstract states [73]. Thus, in each update,

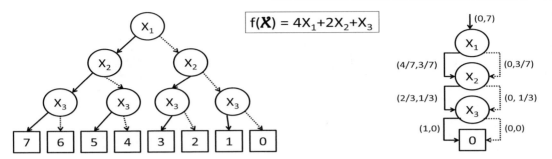

Figure 5.4: ADD and Affine ADD representations for the same function $f = 4X_1 + 2X_2 + X_3$. Notice that the ADD is exponential whereas Affine ADD just takes linear space.

all states that belong to the abstract state with the highest priority are updated. This has similarities with Symbolic RTDP in that a state is generalized to similar states, except that in Structured PS, the generalization is to states that have similar priorities, where in Symbolic RTDP, it is to states with similar values/transition structure.

5.5   OTHER SYMBOLIC REPRESENTATIONS

The biggest advantage of an ADD representation is its ability to encode context-sensitive independence (given a good variable order). On the other hand, a serious disadvantage is its inability to efficiently encode any additive structure. Additive value functions are quite common in MDPs. For example, an MDP may have its reward model such that achieving X_1 gets a reward of 4, achieving X_2 gets 2 and X_3 is worth 1. That is, the reward function is $f(\mathcal{X}) = 4X_1 + 2X_2 + X_3$. An ADD for such an additive function will have an *exponential* number of nodes. Figure 5.4(a) shows the resulting ADD for this function.

ADDs have been extended to Affine ADDs [211] to allow further compressions for the case of additive functions. An Affine ADD is just like an ADD with variables as nodes, and each node with two children, one for the true value and one for the false. The difference is that each child edge is additionally annotated with two constants (c, b). We use c_t and b_t for constants on the true child, and c_f and b_f on the false.

This data structure results in dramatic compressions when the underlying function is additive. Figure 5.4(b) shows an Affine ADD for our function f, which converts an exponential-sized ADD to a linear Affine ADD. The semantics of an Affine ADD is as follows: to get the final value, evaluate the appropriate child subtree, and multiply the evaluation with b and add c.

Example: We evaluate the Affine ADD from Figure 5.4 for the case $X_1 = X_2 = 1$ and $X_3 = 0$. The topmost arrow encodes that we first evaluate the tree (let us call it $V(X_1)$), and then multiply the result by 7 and add 0. Thus, $f = 7V(X_1) + 0$. Since X_1 is true we take the solid branch. We

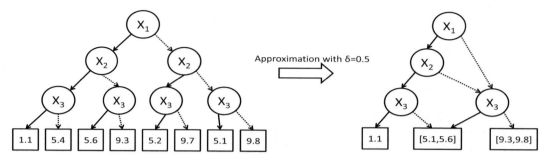

Figure 5.5: Approximating an ADD by merging close-by values can result in a significant reduction in ADD size.

get $V(X_1) = \frac{3}{7}V(X_2) + \frac{4}{7}$. Similarly, $V(X_2) = \frac{1}{3}V(X_3) + \frac{2}{3}$. And since $X_3 = 0$, $V(X_3) = 0 \times 0 + 0 = 0$). Putting this back in we get $V(X_2) = \frac{2}{3}$ and $V(X_1) = \frac{6}{7}$. Thus $f = 7 \times \frac{6}{7} = 6$.□

Affine ADDs, under additional restrictions, can be normalized, i.e., converted to a unique representation for each function. The restrictions are (1) a given variable order and non-repetition of a variable (this is same as in ADDs), (2) for each node, $\min(c_f, c_t) = 0$, (3) for each node $\max(c_f + b_f, c_t + b_t) = 1$, and (4) additional boundary cases when the function's value is 0. Please see [211] for further details.

All the ADD operations, such as 'APPLY,' 'REDUCE,' 'RESTRICT,' 'MARGINALIZE,' and others extend to Affine ADDs. Affine ADD operations are usually slower than ADD algorithms for similar-sized graphs. However, often, the relative compactness of the data structure more than makes up for the additional time requirements.

Since Affine ADDs can perform all the ADD manipulations, it is trivial to implement the MDP algorithms like SPUDD and Symbolic LAO*, to use the Affine ADDs instead of ADDs. These can give even exponential speedups for some problems.

ADDs have been extended to other complex value-function representations too. One of them is First-Order Decision Diagrams (FODDs), which are used in implementing dynamic programming in First-Order MDPs [120]. We briefly discuss these in Section 7.3.1. Recently, a new data structure called Extended ADD (XADD) [209] has been introduced for domains that have both continuous and discrete variables. In XADDs, an internal node can be an inequality over continuous variables. We briefly discuss these in Section 7.1.1.

The use of structured representations in VI is a general theme. Whenever the state space grows too large and flat representations do not scale, there is an attempt to represent the value functions compactly. The VI-based algorithms are most conveniently applied over the representations that are *closed* under the Bellman backup, i.e., V_{n+1} is in the same representation as V_n. We revisit these ideas in MDPs with continuous variables (Section 7.1), since flat representations do not scale for those MDPs.

5.6   APPROXIMATIONS USING SYMBOLIC APPROACHES

Up to this point, we have only described optimal algorithms for solving MDPs. However, at the end of the day, for large problems, finding efficient optimal solutions is infeasible, due to the curse of dimensionality. Therefore, a large body of MDP research focuses on approximations to these algorithms. We discuss several approximation techniques in depth in the next chapter. We close this chapter by describing a simple approximation idea that uses approximate ADDs for dynamic programming.

The leaf nodes in an ADD for $V_n(s)$ represent all the various values that the states can have. We can simply group together those leaves where values are rather close (say, within δ of each other). This will result in an approximation to the original ADD, but could be much smaller in size due to the further application of the 'REDUCE' operator on the approximate ADD.

We store the whole range of values in the leaf node, and use its mean as the value of states mapping to that node. Using such a $\overline{V}_n$ in further computation leads to an approximation of VI known as APRICODD [221]. Figure 5.5 illustrates the potential benefits of this approximation on the size of the resulting ADD.

Similar ideas had been explored before the ADD representations were brought into the MDP literature. These used tree-based value functions [37; 39]. Similar approximation strategies have also been explored with Affine ADDs. Here, not only the leaf nodes but also the similarity of edge constants are used to merge nodes. The resulting algorithm is known as MADCAP [212].

CHAPTER 6

Approximation Algorithms

Although heuristic search methods (especially combined with symbolic representations) enable us to optimally solve larger SSP MDPs than algorithms such as VI and PI, the scalability of all of these approaches is ultimately limited by the same factor — the size of the explicit representation of MDP's optimal policy closed w.r.t. s_0. Recall that heuristic search methods, when used with admissible heuristics, at a minimum, store in memory the entire optimal policy graph rooted at s_0, and potentially other states visited in search of that policy as well. Usually, this takes much less space than a complete policy for the entire state space computed by VI or PI, with structures such as ADDs helping further reduce the memory footprint. Nonetheless, even in the form of a decision diagram, optimal policies closed w.r.t. s_0 can easily exceed the size of available storage. Computing an optimal policy also takes time. While it is a relatively less fundamental issue than the amount of available memory, in many practical applications the agent can plan only for so long before having to take an action. Fortunately, there are several ways of circumventing these limitations if we are willing to settle for a suboptimal, *approximate* solution to the given MDP.

The first two classes of approximation approaches examined in this chapter challenge the assumption that we need to compute a closed specification (Definition 4.2) of an optimal behavior starting at s_0. A closed specification gives an optimal action at any state that the agent *might* reach from the initial state by following this specification. In the meantime, such exhaustiveness may not be necessary in practice. Over its own lifetime, the agent may never visit most of the MDP states it could theoretically end up in via an optimal policy. Consider an MDP for the previously discussed Mars rover example. An optimal policy for this MDP may lead to billions of states from an initial one. Let us assume the figure of one billion for the sake of the example, a very conservative estimate. Now, suppose the rover makes one state transition per second and never visits a state twice. Modern Mars rovers are designed to remain functional for at most several years. Even if its mission lasts for as long as five years, under the above assumptions our rover would visit less than 20% of the possible states. Thus, in this case, much of the resources spent computing an optimal policy closed w.r.t. the initial state would be wasted, i.e., planning *offline* would not pay off.

A more promising approach for such scenarios would be to interleave planning and execution — quickly estimate the best action in the current state *online* by making some approximations, execute the chosen action, replan from the state the system ends up in, and so on. One way to estimate a good action in the current state fast is to *determinize* the MDP, i.e., convert it into a deterministic counterpart, solve the resulting relaxed classical planning problem, and see what actions start good plans from the current state in that relaxation (Section 6.1). The insight behind this idea is that, thanks to advances in classical planning, deterministic planning problems can be solved much more efficiently than equivalently sized probabilistic MDPs. Another class of online approximation algorithms is based on evaluating actions or even entire policies using *Monte Carlo sampling*. These are discussed in Section 6.2.

We will also examine several classes of approximation algorithms that are appropriate if the agent does not have time to stop and think in the middle of executing a policy and has to compute a closed policy entirely in advance, i.e., offline. One approach is to run heuristic search methods guided by *inadmissible* determinization-based heuristics, discussed in Section 6.3. In general, inadmissible heuristics are not necessarily more effective than admissible ones at reducing the memory footprint of heuristic search algorithms; however, those we consider in this chapter in particular often do help save more space than their admissible counterparts. The savings may turn out to be sufficient that heuristic search guided by one of these inadmissible heuristics can find a good policy without running out of space.

In case even heuristic search guided by an inadmissible heuristic runs out of memory, one can use another method, *dimensionality reduction*, which approximates the explicit tabular form of a policy by a more compact representation (Section 6.4). More specifically, the explicit tabular form of a policy can be viewed as a function whose number of parameters is proportional to the number of states that policy can reach from s_0. Dimensionality reduction approaches approximate this function with one that needs many fewer parameters, and hence much less space. An example of such a compact policy encoding is a small set of rules of the type "If it rains outside, take an umbrella" with associated importance weights that tells the agent how to choose an action in any state.

The next type of approaches we will consider in this chapter are the *hierarchical planning* algorithms that solve large MDPs efficiently by breaking them up into a series of smaller ones (Section 6.5). In many settings, a human expert may know what high-level steps a good policy should consist of. For example, a Mars rover mission may involve driving to a particular location, collecting rock samples, and delivering them to a lab at the base. This effectively decomposes the MDP for the entire mission into three subproblems — driving from the base to a certain site, finding the desired samples, and driving back to base. Each of them is significantly smaller and easier to solve than the original problem. If provided with a task decomposition, hierarchical planners take care of solving the set of MDPs corresponding to the steps of high-level plan and "stitching" their solutions together to obtain a policy for the original problem.

Last but not least, we will turn to *hybridized planning* (Section 6.6). The objective of these kinds of offline solvers is to come up with a "reasonable" policy as quickly as possible and then, as time and memory allow, to patch it up in order to endow it with some quality guarantees. Thus, their focus is more on achieving a good anytime performance than on saving memory. More specifically, these planners start by using an optimal heuristic search algorithm. This algorithm rapidly produces a decent policy, but may take a long time to arrive at an optimal solution. Therefore, at some point hybridized planners switch to a suboptimal algorithm that can nonetheless turn the current "half-baked" solution into one with some guarantees, and can do so quickly.

As the final note before we proceed to study the approximation algorithms in detail, we point out that all of them except those that we specifically call out work in the presence of dead ends and thus apply to fSSPUDE MDPs (Definition 4.21). On the one hand, this is not surprising; after all, according to Theorem 4.22, any fSSPUDE MDP can be converted into an SSP_{s_0} MDP. On the other hand, SSP problems derived from fSSPUDE MDPs have a special structure. Therefore, being able to solve them *optimally* with the same techniques as "ordinary" SSP MDPs does not necessarily imply that optimal solutions to these two subtypes of SSP problems are equally easy to *approximate*. In fact, certain determinization-based algorithms have difficulty dealing with dead ends, and we examine reasons for this in Section 6.1.5.

6.1 DETERMINIZATION-BASED TECHNIQUES

Typically, an online algorithm has a very limited amount of time for choosing an action in the current state. For instance, it may be deployed on a delivery robot and may control the robot's motors in order to navigate it to its destination. The robot's controller constantly receives information from the robot's sensors (e.g., people showing up in the robot's path) and has to change the robot's trajectory to account for it. Response time is of essence in such situations, and online algorithms need a fast way to determine a reasonably good action in a state, even if the action is not optimal.

An approach that meets this requirement is to relax the given MDP M into a deterministic problem by assuming that the agent, not nature, chooses actions' outcomes. This amounts to *determinizing* M into a simpler MDP M_d, finding plan(s) from the current state s to the goal in M_d, and use these plan(s) to decide on an action in s in the original MDP M. We have already discussed an example of MDP determinization, the all-outcome determinization (Definition 4.16), and will examine several more in this chapter.

This idea is the basis of all *determinization-based* approaches and relies on a crucial observation — M_d is a classical planning problem, solvable with classical planners. Classical planners solve equivalently sized problems much more quickly than probabilistic solvers, so much so that calling a classical planner many times from different states in M_d is typically still faster than solving M for just one state, s_0, with a dynamic programming-like approach. Thanks to this, the determinization-based online planning algorithms, which we are about to discuss, are among the most efficient approximation schemes for MDPs. Most of them use the deterministic planner called FF (Fast Forward) [112], but other fast solvers such as LAMA [203] would work as well.

Since they critically rely on being able to compute MDP determinizations quickly, and PPDDL-style factored representations make building determinizations particularly fast, all determinization-based algorithms we cover in this chapter assume the MDP to be given in a PPDDL-style factored form. Also, these techniques are primarily aimed at MDPs with a "native" goal state, as opposed to discounted-reward MDPs converted into the goal-oriented form as discussed in Section 2.4.2. The reason for this is that, as we will see in Section 6.1.5, determinization-based algorithms can be viewed as maximizing the *probability* of reaching the goal from the initial state, a proxy optimization criterion for the *expected cost* of reaching the goal. This proxy criterion makes sense for goal-oriented problems but not so much so for the reward-oriented ones formulated as SSP MDPs, since for the latter the probability of reaching the goal is the same for all policies.

6.1.1   FF-REPLAN

FF-Replan is the determinization-based algorithm whose success sparked research into this type of approaches [250]. Before delving into it, we formally define the notion of a *goal trajectory*.

Definition 6.1   Goal Trajectory. A goal trajectory from a state s_0' in a goal-oriented MDP M is a finite sequence $s_0', a_{i_0, j_0}, s_1', a_{i_1, j_1}, s_2', a_{i_2, j_2}, \ldots, s_g'$, where $s_g' \in \mathcal{G}$ and a_{i_k, j_k} is the j_k-th outcome of M's action a_{i_k} s.t. if nature chooses this outcome when a_{i_k} is applied in s_k', the system transitions to state s_{k+1}'.

Thus, a goal trajectory is just a sequence of action outcomes that leads from some state s_0' to a goal state. To emphasize the state transitions, we will denote a goal trajectory as $s_0' : a_{i_0, j_0} \to s_1' : a_{i_1, j_1} \to s_2' : a_{i_2, j_2} \to \ldots \to s_g'$ from now on. We have already encountered goal trajectories when discussing RTDP and LRTDP in Sections 4.4.1 and 4.4.2. In those algorithms, goal trajectories are sampled starting from the MDP's initial state.

Crucially, goal trajectories in an MDP M are closely related to plans to the goal (*goal plans*) in the MDP's all-outcome determinization M_d^a (Definition 4.16). As we discussed in Section 4.6.1 and state in the following theorem, M_d^a is essentially a classical counterpart of M in which each of M's actions' outcomes turns into a separate deterministic action.

Theorem 6.2   Every goal plan from any state s to any goal state s_g in the all-outcome determinization M_d^a of an MDP M is a goal trajectory from s to s_g in M, and every goal trajectory from s to s_g in M is a goal plan from s to s_g in M_d^a.

FF-Replan's central idea is to exploit this property in the following simple way, outlined in Algorithm 6.1. FF-Replan starts by building the all-outcome determinization M_d^a of the given MDP M. Then it finds a deterministic plan from the current state s (initially, $s = s_0$) to the goal in M_d^a using the classical planner FF, and starts executing it *in M*. When some action's probabilistic outcome deviates from the found trajectory, i.e., leads to a state s' that is not the next state in the discovered plan, FF-Replan replans to find a deterministic goal plan from s' and repeats the cycle. It keeps doing so until a goal state is reached.

More specifically, each discovered plan in M_d^a has the form $s_0' : a_{i_0, j_0} \to s_1' : a_{i_1, j_1} \to s_2' : a_{i_2, j_2} \to \ldots \to s_g'$ — exactly the same form as the corresponding goal trajectory in M. Executing this plan in M means trying to apply the action sequence $a_{i_0}, a_{i_1}, \ldots$ starting at the current state s_0'. An important implication of Theorem 6.2 is that doing so may eventually bring the agent to the goal with non-zero probability. Note, however, that attempting to execute the above action sequence in M may also at some point take the system to a state not anticipated by this deterministic plan. For instance, applying a_{i_0} in s_0' will not necessarily cause a transition to s_1', because nature may choose an outcome of a_{i_0} different from a_{i_0, j_0}. If this happens, FF-Replan finds a plan in M_d^a from wherever the system ended up to the goal and makes the agent follow it instead. In short, FF-Replan's strategy consists of finding goal trajectories via classical planning in accordance with Theorem 6.2 and guiding the agent along these trajectories toward the goal.

Algorithm 6.1: FF-Replan

  1 $M_d^a \leftarrow$ all-outcome determinization of M

  2 $s \leftarrow s_0$

  3 // Find a plan from s to a goal state $s_g' \in \mathcal{G}$ in M_d^a using the FF planner

  4 $currDetPlan \xleftarrow{FF} [s : a_{i_0, j_0} \rightarrow s_1' : a_{i_1, j_1} \rightarrow s_2' : a_{i_2, j_2} \rightarrow \ldots \rightarrow s_g']$

  5 $k \leftarrow 0$

  6 $s \leftarrow$ execute action a_{i_k} in s in the original MDP M

  7 **while** $s \notin \mathcal{G}$ **do**

  8   **if** $s \neq s_k'$ **then**

  9   // Find a new plan from s to a goal state $s_g' \in \mathcal{G}$ in M_d^a using FF planner

 10         $currDetPlan \xleftarrow{FF} [s : a_{i_0, j_0} \rightarrow s_1' : a_{i_1, j_1} \rightarrow s_2' : a_{i_2, j_2} \rightarrow \ldots \rightarrow s_g']$

 11         $k \leftarrow 0$

 12   **end**

 13   **else**

 14         $k \leftarrow k + 1$

 15   **end**

 16      $s \leftarrow$ execute action a_{i_k} in s in the original MDP M

 17 **end**

An early version of FF-Replan [250] uses the *most likely outcome determinization* instead of the all-outcome one:

Definition 6.3   Most-Likely-Outcome Determinization. The *most-likely-outcome determinization* of an MDP $M = \langle \mathcal{S}, \mathcal{A}, \mathcal{T}, \mathcal{C}, \mathcal{G} \rangle$ is a deterministic MDP $M_d^m = \langle \mathcal{S}, \mathcal{A}', \mathcal{T}', \mathcal{C}', \mathcal{G} \rangle$ s.t. for each triplet $s \in \mathcal{S}, a \in \mathcal{A}, s' = \text{argmax}_{s' \in \mathcal{S}} \mathcal{T}(s, a, s')$, the set $\mathcal{A}'$ contains an action a' for which $\mathcal{T}'(s, a', s') = 1$ and $\mathcal{C}'(s, a', s') = \mathcal{C}(s, a, s')$.

Put simply, for each state s and action a, M_d^m only has an action corresponding to the most likely outcome of a in s. As a result, M_d^m is typically smaller and faster to solve than M_d^a. At the same time, it may be impossible to get to the goal from a given state using only the most likely outcome of each action. In states where this is the case, the M_d^m-based flavor of FF-Replan fails, making it less robust than the version that uses the all-outcome determinization M_d^a.

When and why does FF-Replan excel? At a high level, FF-Replan performs very well in online settings in MDPs where deviations from the chosen linear plan are not too costly. As an example, imagine navigating a robot whose actuators have some small probabilistic output noise. When ordered to drive straight, this robot may follow a slightly curved path, and when ordered to turn by x degrees it may actually turn by $x \pm \epsilon$ degrees, ϵ being a small deviation from the desired angle. Provided that the controller monitors the robot's behavior and issues commands to correct it frequently enough relative to the robot's movement speed, the noise in the motors will not have any fatal effects, although it may make the robot's trajectory slightly erratic. In situations like this, FF-Replan is a natural choice. Indeed, in order to plan a route from a source to a destination, it makes sense to *pretend* that the controller can choose the desired effect of an action, let it construct a plan accordingly, and then simply adjust the robot's trajectory by replanning if the robot deviates from the current plan. Another reason why FF-Replan is ideal for these scenarios is its speed — the FF planner at the core of FF-Replan is fast even on determinizations of very large MDPs.

At the same time, FF-Replan has major trouble with MDPs where trying to execute a goal plan from their determinization results in catastrophic consequences with a high probability (e.g., causes a transition to a dead end). We discuss this phenomenon in detail in Section 6.1.5. In a sense, the determinization-based algorithms we turn to next are attempts to remedy FF-Replan's shortfalls.

6.1.2   FF-HINDSIGHT

FF-Replan's main weakness lies in its failure to anticipate possible deviations from the linear plan found by a deterministic planner. In turn, the linear plan is constructed without taking into account how likely the actions' outcomes that form this plan are. Hence, FF-Replan's chosen course of action may be successful only with a lot of luck, and deviations from it may be very costly.

A natural way to mitigate this drawback is to check in advance if the goal can be reached cheaply even if actions' outcomes happen to deviate from the original plan. More specifically, for each action in the current state, it would be good to find *several* goal plans that start with different outcomes of this action and, in general, use different outcomes of the MDP's actions to get to the goal. We could then estimate the Q-value of each action in the state as the average of the costs of these plans. Intuitively, doing so would let us "in hindsight" choose an action that, in expectation, starts us off on a cheap path to the goal.

This basic idea is the foundation of an online planner called FF-Hindsight [251]. FF-Hindsight takes two parameters — T, the number of lookahead steps, and W, the number of attempts to find deterministic plans per action. To select an action in the current state s, for each action in s it samples and tries to solve W *futures*. A future is a non-stationary determinization of the MDP, a classical planning problem where at each time step $t = 1, \ldots, T$ only one (randomly sampled) outcome of each action is available for use as the next step in the plan. Sampling a future means sampling, for each probabilistic action a and time step $t = 1, \ldots, T$, the outcome of a that may be used for constructing the t-th step of a goal plan. Each outcome is chosen according to its probability in the original MDP. Solving a future amounts to finding a plan, with the help of the aforementioned FF planner, that reaches the goal from s within at most T steps using only the actions sampled in this future.

FF-Hindsight samples W futures of each possible action at the current time step $t = 1$. The purpose of sampling several futures for each action in the current state is to simulate various situations in which the system may end up in $t = 1, \ldots, T$ time steps if this action is chosen now, and to see if the agent can reach the goal cheaply from those situations. Once FF-Hindsight solves W futures for an action a in s, it approximates $Q^*(s, a)$ as the average of the found plans' costs (and dead-end penalties for those futures that it could not solve). It then selects the action with the lowest such average, and, after the agent executes the chosen action in s, repeats the process in the state where the agent transitions.

There is an intuitive relationship between the basic ideas of FF-Hindsight and FF-Replan. In the current state, FF-Replan chooses an action that *under "ideal" circumstances* starts the agent off on a path that will bring the agent to the goal in a cheap way. FF-Hindsight chooses an action that *in many circumstances* starts the agent off on a path that will bring the agent to the goal in a cheap way. On many nontrivial problems, FF-Replan's approach is overly optimistic and may lead to a disaster, since the agent may easily deviate from the plan into a dead end. Contrariwise, FF-Hindsight copes with many such problems well if the number of futures W that its decisions are based on is large enough and the goal can be reached from s_0 within T steps with probability 1. On the other hand, computationally, FF-Hindsight can be much more expensive. It may need to invoke a classical planner $W \cdot |\mathcal{A}|$ times for each state the agent visits. A series of improvements [249] to the basic FF-Hindsight algorithm have made it more efficient by adding action selection heuristics and other modifications. Nonetheless, as for many other pairs of planners, the choice between FF-Hindsight and FF-Replan is often one between solution quality and speed.

6.1.3   RFF

RFF (Robust FF) [228] is another approach for mitigating FF-Replan's lack of contingency reasoning. Recall that FF-Replan essentially finds a partial policy that will most likely need replanning due to unexpected action outcomes. In contrast, the main idea of RFF is to incrementally build up a partial policy that will probably *not* need replanning. In particular, with every partial policy that prescribes an action for a state s, RFF associates the *failure probability* of that policy w.r.t. s. It denotes the probability that, by following this partial policy starting at s, the agent will eventually reach a state for which the policy does not specify any action, i.e., for which the policy *fails*. By analogy with LAO* (Section 4.3.1), the states where the current policy does not know how to behave are called the *fringe* states. RFF's ultimate objective is to refine the partial policy until its failure probability (the probability of reaching a fringe state) falls below some small value given by the input parameter ϵ. This objective gives RFF its name — unlike FF-Replan, RFF strives to be robust to deviations from the original linear plan, whose failure probability is typically very high.

RFF has an offline and an online stage. In the offline stage, as long as it has time, it attempts to compute a partial policy with $P(failure) < \epsilon$ at the initial state s_0. In the online stage, it executes that partial policy and, whenever it happens to run into fringe states, computes partial policies with a low failure probability for them as well. It uses the same algorithm for computing partial policies in both stages.

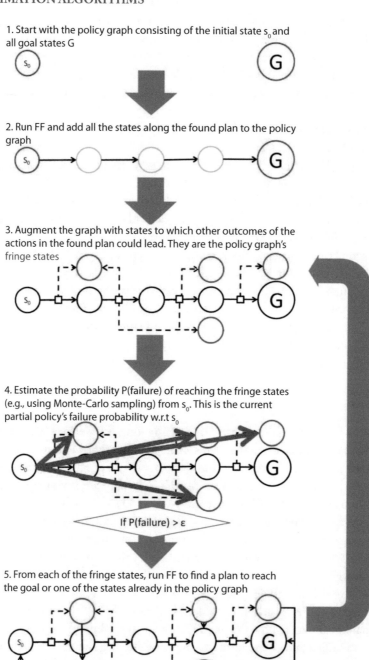

Figure 6.1: RFF: a scheme for computing a partial policy with a low failure probability.

The high-level scheme of RFF's policy computation is presented in Figure 6.1. Without loss of generality, we will explain its operation in the offline stage, when it constructs a policy tree rooted at s_0. The algorithm consists of five basic steps. First, RFF initializes the policy graph to include only s_0 and the goal states. Next, in step 2, it finds a plan from s_0 to a goal state in a determinization of the given MDP, and adds this plan (essentially, a goal trajectory) to the policy graph. So far, RFF behaves the same way as FF-Replan— it looks for a deterministic plan and, for every state s in this plan, makes its own policy choose the action a that the discovered plan uses in s. However, starting from step 3, RFF does something different. Instead of attempting to execute the discovered linear plan, it augments the partial policy graph with edges corresponding to *all* possible outcomes of actions used in that linear plan. It also puts the states these outcomes lead to into the policy graph, unless the graph already contains these states. These newly added states form the fringe — RFF does not yet know what actions to choose in them. Essentially, in step 3, RFF explores where the agent may end up in if it deviates from the current partial policy. In step 4, RFF evaluates the current policy's *P(failure)* w.r.t. s_0 by performing Monte Carlo simulations. If *P(failure)* turns out to be less than ϵ, RFF proceeds to executing the policy. Otherwise, in step 5, it computes goal plans from all the fringe states and goes back to step 3 to account for the possibility of deviation from these plans.

The performance of RFF depends on several choices:

- The choice of determinization. Besides the all-outcome determinization, RFF may benefit from using the most-likely-outcome determinization (Definition 6.3), since solving the latter is usually faster. At the same time, as mentioned in the discussion of FF-Replan, the most-likely-outcome determinization has no plans corresponding to many positive-probability goal trajectories. As a price for extra efficiency, this may lead RFF to give up in some states that are perfectly solvable in the all-outcome determinization.

- The choice of goal states during replanning. When looking for goal plans from the fringe states in step 5, an obvious decision is to aim directly for the goal states. However, it may not be the best one, since goal states may be far away. A more promising choice is to try to find plans to states *already in the partial policy graph*. Typically, there are more such states than goals, and finding a plan to any of them automatically gives a plan to a goal state.

- The option of performing stochastic optimization on the partial policy. As described, RFF ignores the probabilistic structure of the problem. In choosing actions, it effectively always follows the advice of a deterministic planner. If actions participating in deterministic plans are very suboptimal (e.g., lead to a dead end with high probability when dead ends are avoidable), RFF makes no attempt to pick different ones. A modification that enables RFF to select actions more intelligently is to perform Bellman backups on the states in the partial policy graph after it is expanded in step 3. This is another similarity between RFF and LAO*; the latter resorts to the same strategy in order to propagate state values in a decision-theoretic way. Bellman backups alter RFF's partial policy graph; in particular, they can help the agent avoid dead ends.

- The choice of the failure probability threshold ϵ. The value of ϵ has a crucial influence on RFF's speed. If ϵ is low, RFF will try to make the policy graph as complete as possible, essentially operating like an offline algorithm. If ϵ is close to 1, however, RFF will be as oblivious of possible deviations from the constructed policy as FF-Replan, but also as fast. The optimum lies somewhere in the middle. FF-Replan easily deviates from the plans it finds, often causing the agent to revisit some states on its way to the goal. Every time the agent revisits a state as a result of a deviation, FF-Replan has to look for a goal plan from that state yet another time, despite having done that before. If the ϵ parameter is tuned well, RFF avoids this wastefulness by memoizing the already-found plans and organizing them into a policy tree, but at the same time has a small memory footprint.

RFF is a special instance of a more general strategy known as *incremental contingency planning* [75]. Like RFF, approaches of this type first come up with a policy backbone and then decide where to add branches to it to make the policy specification more complete. RFF chooses to add branches for all possible outcomes of actions used in the already constructed part of the policy. However, many other strategies are also possible, e.g., augmenting the existing partial policy only at the most likely action outcomes that have not been explored yet.

6.1.4   HMDPP

The approaches of FF-Replan, FF-Hindsight, and RFF are largely motivated by the question: how should we deal with possible deviations from deterministic goal plans? FF-Replan simply follows the plan produced by a deterministic planner without trying to anticipate deviations, while FF-Hindsight and RFF attempt to assess the consequences of the deviations in advance. There is also a non-obvious third way, to trick a deterministic planner into generating action sequences with a low probability of deviation to start with. This is the underlying idea of the HMDPP planner [126]. Like other approximate solvers considered in this section, HMDPP is also largely guided by plans produced by a deterministic solver. However, it constructs the determinization on which this solver operates in such a way that the cost of a plan in the determinization takes into account the probability of deviating from it in the original MDP. This forces plans with a high probability of deviation to look expensive and therefore unattractive to the deterministic planner, encouraging the planner to generate "probabilistically reliable" plans. The key to HMDPP's strategy is the *self-loop determinization*:

Definition 6.4   Self-loop Determinization. The self-loop determinization of an MDP $M = \langle \mathcal{S}, \mathcal{A}, \mathcal{T}, \mathcal{C}, \mathcal{G} \rangle$ is the deterministic MDP $M_d^{sl} = \langle \mathcal{S}, \mathcal{A}', \mathcal{T}', \mathcal{C}', \mathcal{G} \rangle$ s.t. for each triplet $s, s' \in \mathcal{S}$, $a \in \mathcal{A}$, for which $\mathcal{T}(s, a, s') > 0$, the set $\mathcal{A}'$ contains an action a' for which $\mathcal{T}'(s, a', s') = 1$ and $\mathcal{C}'(s, a', s') = \frac{\mathcal{C}(s, a, s')}{\mathcal{T}(s, a, s')}$.

The self-loop determinization differs from the all-outcome counterpart in only one aspect, the costs of its actions. The cost of each action in the former is divided by the probability of the corresponding outcome in the original MDP. To see why this makes sense, imagine that the agent is in a state s and is about to execute some action a that is part of a deterministic plan it is trying to follow. Since the plan is deterministic, only one outcome of a, s', is actually part of the plan, while others mean deviating from it. Thus, when executing a, the agent is hoping for a specific outcome to happen. Now, suppose that whenever the desired outcome does *not* happen, the agent simply stays in s, as opposed to transitioning to some other unanticipated state. In this case, deviating from the intended plan becomes impossible. Following it through to the goal is only a matter of patience and cost — if at some step of the plan the desired outcome of the appropriate action (for instance, a) is not realized, all the agent has to do is try the action again and again until the intended transition finally happens. However, every time the agent retries a, it pays the cost of a (according to the original MDP). Moreover, the less likely the intended outcome of a is, the more times in expectation the controller has to retry it. This process is similar to buying lottery tickets until one of them wins — the smaller the probability of winning, the more money one needs to spend on getting the tickets. What is the total expected cost of getting the intended outcome s' of a in s ("the winning ticket")? It is $\frac{C(s,a,s')}{T(s,a,s')}$ (the price of a ticket divided by the probability of winning), which is exactly the cost of a in s in the self-loop determinization M_d^{sl}.

Thus, the self-loop determinization makes the cost of an action be the expected cost of getting the desired outcome of the action (in the hypothetical situation when repeating the action is possible) by retrying the action over and over again. For example, if an action has only one possible outcome in the original MDP (i.e., the action is deterministic), its costs in the original MDP and in the self-loop determinization coincide. However, as the probability of an action's outcome tends to 0, the cost of the action corresponding to that outcome in M_d^{sl} tends to infinity. Thus, goal trajectories that are expensive or easy to deviate from in M are expensive goal plans in M_d^{sl}.

The self-loop determinization gives us a convenient non-naive way of estimating the value of any successor s' of s under any action a (and hence, ultimately, of estimating the Q-value of a in s) as the cost of a goal plan originating in s' in M_d^{sl}. We could find such a plan with the classical FF planner, as FF-Replan does. However, HMDPP uses a cheaper way; it approximates the plan's cost without computing the plan itself by applying the h_{add} heuristic, covered in Section 6.3.1, to s'. The h_{add} heuristic is inadmissible (which is why we discuss it later, together with other inadmissible heuristics), but gives highly informative estimates of states' values.

Let us denote the h_{add} heuristic applied in M_d^{sl} as h_{add}^{sl}. In the current state s, HMDPP estimates the Q-value of each action a by applying the Q-value formula (Definition 3.7) to the heuristic value function h_{add}^{sl}:

$$Q_{add}^{sl}(s, a) = \sum_{s' \in \mathcal{S}} T(s, a, s')[C(s, a, s') + h_{add}^{sl}(s')] \tag{6.1}$$

HMDPP also assesses the Q-value of each action in the current state using a pattern database heuristic, h_{pdb} [126], producing an estimate $Q_{pdb}(s, a)$. This heuristic helps HMDPP recognize dead-end states and consequently avoid them. To select an action in s, HMDPP ranks all actions applicable in s according to a combination of the $Q_{add}^{sl}(s, a)$ and $Q_{pdb}(s, a)$ values, and chooses an action a^* that minimizes this combination. The agent executes a^*, transitions to another state, and calls HMDPP to repeat the procedure.

Thanks to its strategy, HMDPP's has been shown to outperform RFF on a number of hard problems [46].

6.1.5   DETERMINIZATION-BASED APPROXIMATIONS AND DEAD ENDS

Compared to other approximation frameworks that are discussed later in this chapter, getting determinization-based algorithms (which ignore a large part of probabilistic information in an MDP) to assess the risk of running into a dead end has proved particularly challenging. In this section, we explore this issue in some detail.

When FF-Replan, the first of determinization-based algorithms, made its debut, it was tested mostly on SSP_{s_0} MDPs without dead ends and vastly outperformed its competition on these types of problems. Although the expected cost of its policies could be far from optimal, its use of a deterministic planner made sure that it could produce these policies quickly and required little memory. However, on MDPs with dead ends, FF-Replan immediately started to show a vice. FF-Replan's strategy of ignoring possible deviations from the linear plans it picks meant that it would often run into a dead-end state when nature chose an unforeseen outcome of one of the executed actions. An example of such a setting is controlling an airplane. Sometimes, the shortest path to the airplane's destination lies through a storm. Flying through a storm has a much higher risk of air disaster than flying around it, even though the latter may be longer and may have a higher cost in terms of fuel. In an MDP describing such a scenario, FF-Replan will try to follow the sequence of actions recommended by FF. In turn, FF usually prefers shorter action sequences, since it indirectly attempts to minimize the number of steps in the deterministic plan. Thus, FF-Replan would likely direct the airplane into the storm — an option with potentially catastrophic consequences. This behavior made researchers realize that, in designing determinization-based approximation techniques for realistic scenarios, one needed to pay special care to these techniques' ability to avoid dead ends.

More specifically, recall that in fSSPUDE MDPs, dead ends are merely states with a very high penalty. Thus, a policy for these MDPs may have a suboptimally high expected cost for one of two reasons (or both) — either its trajectories are costly in expectation, or it has a high probability of leading to a dead end. In many scenarios, reducing solution quality due to the former cause is the lesser evil. For instance, returning to the example of an airplane trying to navigate around a storm, we would much rather prefer a policy that is suboptimal in terms of fuel cost than one that runs an increased risk of causing a disaster by vectoring the airplane into the storm. In other words, for an fSSPUDE MDP, we would like approximation techniques to produce policies that may be subpar according to the expected cost criterion but are nearly optimal in terms of their probability of reaching the goal. Policies satisfying the latter criterion can be understood as solutions to the $MAXPROB_{s_0}$ MDP [137] derived from the given fSSPUDE MDP:

Definition 6.5 $MAXPROB_{s_0}$ MDP. A $MAXPROB_{s_0}$ MDP, denoted as $MAXPROB_{s_0}$, is a tuple $\langle \mathcal{S}, \mathcal{A}, \mathcal{T}, \mathcal{R}, \mathcal{G}, s_0 \rangle$, where $\mathcal{S}, \mathcal{A}, \mathcal{T}, \mathcal{G}$, and s_0 are as in the definition of an SSP_{s_0} MDP (Definition

4.3), and the reward function $\mathcal{R}$ obeys two conditions:

- For all pairs of states $s, s' \notin \mathcal{G}$ and all actions $a \in \mathcal{A}$, $\mathcal{R}(s, a, s') = 0$.

- For all pairs of states $s \notin \mathcal{G}$, $s_g \in \mathcal{G}$ and all actions $a \in \mathcal{A}$, $\mathcal{R}(s, a, s_g) = 1$.

The *derived MAXPROB$_{s_0}$ MDP* for an fSSPUDE MDP $\langle \mathcal{S}, \mathcal{A}, \mathcal{T}, \mathcal{C}, \mathcal{G}, \mathcal{P}, s_0 \rangle$ is a tuple $\langle \mathcal{S}, \mathcal{A}, \mathcal{T}, \mathcal{R}, \mathcal{G}, s_0 \rangle$, where $\mathcal{R}$ obeys the two conditions above.

Since MAXPROB$_{s_0}$ MDPs are reward-based, an optimal solution in them *maximizes* the expected reward of getting to the goal. Note, however, that, although viewing policies in MAXPROB$_{s_0}$ MDPs as maximizing reward is mathematically correct, there is a more intuitive interpretation of the MAXPROB optimization criterion. MAXPROB$_{s_0}$ MDPs have a very special structure — the only transitions in them that yield a non-zero reward are those from non-goal states to goal states, and all such transitions yield the reward of 1. This means that the optimal value function in a MAXPROB$_{s_0}$ MDP for every state gives *the maximum probability of reaching the goal from that state*. For example, in a MAXPROB$_{s_0}$ MDP derived from an fSSPUDE MDP, $V^*(s) = 1$ for states for which there exists some proper policy, and $V^*(s) = 0$ for dead ends.

The evolution of determinization-based MDP algorithms starting from FF-Replan can be viewed as a series of attempts to improve the quality of solutions they produce on MAXPROB$_{s_0}$ MDPs derived from fSSPUDE problems. As illustrated above, the probability of reaching the goal is often a more informative indicator of the approximate solution quality for fSSPUDE MDPs than expected cost per se. Therefore, developing determinization-based techniques to optimize derived MAXPROB$_{s_0}$ MDPs is an attractive research objective — achieving it would give us scalable (as FF-Replan has proven itself to be) tools for obtaining meaningful approximations to a hard class of problems. We now turn to examining how well the determinization-based approaches described in Section 6.1 meet this objective:

- **FF-Replan**, as already mentioned, does rather poorly in terms of solving MAXPROB$_{s_0}$ MDPs. An informal subtype of MAXPROB$_{s_0}$ MDPs on which FF-Replan-produced policies are especially prone to hit a dead end is *probabilistically interesting MDPs* [154]. Roughly, these are goal-oriented MDPs with dead ends where different outcomes of an action can lead to states with different probabilities of reaching the goal. It is easy to see why such MDPs confound FF-Replan. FF-Replan always tries to execute a linear plan found by a deterministic plan; since actions in this plan typically have several outcomes, FF-Replan can be expected to easily deviate from it. In general, the deviations are not necessarily a problem. However, in probabilistically interesting MDPs they may lead the agent to a state, reaching the goal from which has a low probability under any policy. On some sets of pathological MDP examples, FF-Replan does arbitrarily poorly in terms of the probability of reaching the goal as the MDP size increases [154].

- **FF-Hindsight** considers several trajectories when evaluating a state, and therefore is much better at avoiding dead ends than FF-Replan. However, in order to reliably assess the danger on ending up in a dead end from a given state, it has to sample many deterministic futures, exacting a steep price in the speed of computing a solution.

- **RFF**, although aware of dead ends' existence, does little to avoid them. When, as a result of expansion, its policy graph includes a dead end, the basic RFF version does not modify the policy to avoid the dead end. Rather, it simply excludes the dead end from the policy graph and marks it in order to prevent its re-inclusion in the future. A more sophisticated RFF version performs policy optimization [228] on the solution graph using Bellman backups, which improves its ability to eschew dead ends when possible.

- **HMDPP** detects dead-end states explicitly with the help of a dedicated heuristic. Its approach appears to strike a good balance between the quality of a solution according to the MAXPROB criterion and the efficiency of finding it.

Our discussion of approximating solutions to $MAXPROB_{s_0}$ MDPs with determinization-based algorithms would be incomplete without considering a natural alternative: why not solve MAXPROB problems *optimally*? After all, perhaps solving MAXPROB is easy enough that it can be done reasonably efficiently without resorting to tricks such as determinization? Unfortunately, too little is known about $MAXPROB_{s_0}$ MDPs to answer these questions conclusively. *MAXPROB* is known not to be a subclass of *SSP* with an initial state [137], since in $MAXPROB_{s_0}$ MDPs, improper policies do not accumulate an infinite cost, as the *SSP* definition requires. Rather, $MAXPROB_{s_0}$ is a subclass of *GSSP* [137], a type of MDPs with complicated mathematical properties briefly covered in Chapter 7. The most efficient currently known method of optimally solving $MAXPROB_{s_0}$ MDPs is heuristic search, although of a more sophisticated kind [137] than presented in Chapter 4. However, at present its effectiveness is limited by the lack of admissible heuristics for MAXPROB problems. The only nontrivial (i.e., more involved than setting the value of every state to 1) such heuristic known so far is SixthSense [133], which soundly identifies dead ends in the MDP's state space and sets $h_{6S}(s) = 0$ for them. Curiously, computing SixthSense itself heavily relies on domain determinization. We revisit it in some more detail in Section 7.4.3.

6.2   SAMPLING-BASED TECHNIQUES

Both the optimal and the approximate MDP algorithms discussed so far work well as long as the number of outcomes of each action in different states is, on average, small. For a factored MDP, "small" means constant or polynomial in the number of state variables. There are plenty of scenarios for which this is not the case. To illustrate them, we use a toy problem named Sysadmin, introduced in Section 2.5.3. Recall that it involves a network of n servers, in which every running server has some probability of going down and every server that is down has some probability of restarting at every time step. There are n binary variables indicating the statuses of all servers, and at each time step the system can transition to virtually any of the 2^n states with a positive probability. Some MDP specification languages, e.g., RDDL [204], which we saw in the same section, can easily describe such MDPs in a compact form (polynomial in the number of state variables).

What happens to the algorithms we have explored in this book so far if they are run on an MDP with an exponentially sized transition function, as above? Let us divide these techniques into two (non-mutually exclusive) groups, those that compute Q-values via Definition 3.7 (e.g., RTDP) and those that use determinizations (e.g., FF-Replan). Computing a Q-value requires iterating over all of an action's successors, whose number for the MDPs we are considering is exponential in the problem description size. Thus, the first group of methods will need to perform a prohibitively expensive operation *for every state transition*, rendering them impractical.[1] The second group of methods does not fare much better. In the presence of a large transition function, their preprocessing step, generating the determinization, requires an exponential effort as well. Thus, all the machinery we have seen seems defeated by such MDPs.

6.2.1   UCT

UCT [128] is a planning algorithm that can successfully cope even with exponential transition functions. Exponential transition functions plague other MDP solvers ultimately because these solvers attempt to enumerate the domain of the transition function for various state-action pairs, e.g., when summing over the transition probabilities or when generating action outcomes to build determinizations. As a consequence, they need the transition function to be explicitly known and efficiently enumerable. Instead, being a Monte Carlo planning technique [89], UCT only needs to be able to efficiently sample from the transition function and the cost function. This is often possible using an environment simulator even when the transition function itself is exponentially sized. Moreover, the fact that UCT only needs access to the transition and cost functions in the form of a simulator implies that UCT does *not* need to know the transition probabilities (and action costs) explicitly.

As Algorithm 6.2 shows, UCT works in a similar way to RTDP. It samples a number of trajectories, or *rollouts* in UCT terminology, through the state space, and updates Q-value approximations for the state-action pairs the trajectories visit. The rollouts have the length of at most T_{max}, a user-specified cutoff, as is common for RTDP as well. However, the way UCT selects an action in a state, and the way it updates Q-value approximations is somewhat different from RTDP. For each state s, UCT maintains a counter n_s of the number of times state s has been visited by the algorithm. The n_s counter is incremented every time a rollout passes through s. For each state-action pair $\langle s, a \rangle$, UCT maintains a counter $n_{s,a}$ of how many times UCT selected action a when visiting state s. Clearly, for each s, $n_s = \sum_{a \in \mathcal{A}} n_{s,a}$. Finally, for each $\langle s, a \rangle$, UCT keeps track of $\hat{Q}(s, a)$, an approximation of the Q-value of a in s equal to the average reward accumulated by past rollouts after visiting s and choosing a in s (line 30). In every state s, UCT selects an action

$$a' = \operatorname*{argmin}_{a \in \mathcal{A}} \left\{ \hat{Q}(s, a) - C \sqrt{\frac{\ln(n_s)}{n_{s,a}}} \right\} \qquad (6.2)$$

---

[1]The use of ADDs may alleviate the problem of computing Q-values in some, though not all, such problems.

Algorithm 6.2: UCT

1 **while** *there is time left* **do**
2      $s \leftarrow$ current state
3      $cumulativeCost[0] \leftarrow 0$
4      $maxNumSteps \leftarrow 0$
5   // Sample a rollout of length at most T_{max}, as specified by the user
6   **for** $i = 1$ *through* T_{max} **do**
7   **if** *s has not been visited before* **then**
8              $n_s \leftarrow 0$
9   **for** *all* $a \in \mathcal{A}$ **do**
10                  $n_{s,a} \leftarrow 0$
11                  $\hat{Q}(s,a) \leftarrow 0$
12   **end**
13   **end**
14          $maxNumSteps \leftarrow i$
15   // $C \geq 0$ is a user-specified weight of the exploration term
16          $a' \leftarrow \mathrm{argmin}_{a \in \mathcal{A}} \left\{ \hat{Q}(s,a) - C \sqrt{\frac{\ln(n_s)}{n_{s,a}}} \right\}$
17          $s' \leftarrow$ execute action a' in s
18   // $cumulativeCost[i]$ is the cost incurred during the first i steps of the current rollout
19          $cumulativeCost[i] \leftarrow cumulativeCost[i-1] + \mathcal{C}(s,a,s')$
20          $s_i \leftarrow s$
21          $a_i \leftarrow a'$
22          $s \leftarrow s'$
23   **if** $s \in \mathcal{G}$ **then**
24   break
25   **end**
26   **end**
27   **for** $i = 1$ *through* $maxNumSteps$ **do**
28   // Update the average $\hat{Q}(s_i, a_i)$ with the total cost incurred in the current rollout
29   // after visiting the state-action pair $\langle s_i, a_i \rangle$
30          $\hat{Q}(s_i, a_i) \leftarrow \frac{n_{s_i,a_i} \hat{Q}(s_i,a_i) + (cumulativeCost[maxNumSteps] - cumulativeCost[i-1])}{n_{s_i,a_i} + 1}$
31          $n_{s_i} \leftarrow n_{s_i} + 1$
32          $n_{s_i,a_i} \leftarrow n_{s_i,a_i} + 1$
33   **end**
34 **end**
35 return $\mathrm{argmin}_{a \in \mathcal{A}} \hat{Q}(\text{current state}, a)$

The algorithm then samples an outcome of a' and continues the trial.

     UCT effectively selects an action in each state based on a combination of two characteristics — an approximation of the action's Q-value ($\hat{Q}(s,a)$) and a measure of how well-explored the action in this state is ($C \sqrt{\frac{\ln(n_s)}{n_{s,a}}}$). Intuitively, UCT can never be "sure" of how well $\hat{Q}(s,a)$ approximates the quality of an action because $\hat{Q}(s,a)$ is computed via sampling and may fail to take into account some of a's possible effects. Therefore, there is always the danger of $\hat{Q}(s, a)$ failing to reflect some particularly good or bad outcome of a, which could affect the ordering of actions in a state by quality. In other words, UCT needs to always be aware that some actions are not explored well enough. One way to implement this insight is to have UCT from time to time try an action that currently seems suboptimal but has not been chosen for a long time. This is exactly what the exploration term $C\sqrt{\frac{\ln(n_s)}{n_{s,a}}}$ does for UCT. If a state s has been visited many times ($\ln(n_s)$ is large) but a in s has been tried few times ($n_{s,a}$ is small), the exploration term is large and eventually forces UCT to choose a. The constant $C \geq 0$ is a parameter regulating the relative weight of the exploration term and the Q-value approximation [128]. The value of C greatly affects UCT's performance. Setting it too low causes UCT to under-explore state-action pairs. Setting it too high slows down UCT's convergence, since the algorithm spends too much time exploring suboptimal state-action pairs.

The exploration term may appear to make UCT's action selection mechanism somewhat arbitrary. When UCT starts exploring the MDP, this is so indeed; the exploration term encourages UCT to try a lot of actions that, purely based on $\hat{Q}(s, a)$, look suboptimal. However, notice: as the denominator $n_{s,a}$ grows, it is harder and harder for the numerator $\ln(n_s)$ to "keep up the pace," i.e., the growth of the exploration term for $\langle s, a \rangle$ slows down. That is, the more times UCT visits a state-action pair, the smaller this pair's exploration term and the bigger the role of $\hat{Q}(s, a)$ in deciding whether a is selected in s. The longer the algorithm runs, the more its action selection strategy starts resembling RTDP's.

UCT has no termination criterion with solution quality guarantees, but the following result holds regarding its convergence.

Theorem 6.6 In a finite-horizon MDP where all action costs have been scaled to lie in the [0,1] interval and for each augmented state (s, t) the constant C of the exploration term is set to t, the probability of UCT selecting a suboptimal action in the initial state s_0 converges to zero at a polynomial rate as the number of rollouts goes to infinity [128].

Note that this result pertains to finite-horizon MDPs and does not directly apply to the goal-oriented setting. However, if for an SSP_{s_0} MDP it is known that there is a policy that reaches the goal from s_0 within H steps with probability 1, one can set the T_{max} parameter (line 6) to H and expect that UCT will approach a near-optimal policy as the theorem describes.

In an online setting, UCT can be used by giving it some amount of time to select an action in the current state and making it return an action when the time is up. The amount of planning time to allocate for good results is problem-dependent and needs to be determined experimentally. UCT can also greatly benefit from a good initialization of its Q-value approximations $\hat{Q}(s, a)$. Well-tuned UCT versions are responsible for several recent, as of 2012, advances in computer game playing (e.g., bringing AI to a new level in 9×9 Go [93]). It is also the basis of a successful planner for finite-horizon MDPs called PROST [124; 204]. The family of Monte Carlo planning algorithms to which UCT belongs is a vast research area, most of which is beyond the scope of this book, and we encourage the reader to use external sources to explore it [89].

6.3 HEURISTIC SEARCH WITH INADMISSIBLE HEURISTICS

The online planning techniques are applicable when the agent can afford some planning time during policy execution. This may not be so, e.g., in scenarios where the agent is trying to play some fast-paced game such as soccer. In these cases, it may be worth trying to use a heuristic search method with an inadmissible heuristic of the kind discussed in this section to learn a small suboptimal policy closed w.r.t. s_0 before taking any action.

Indeed, many MDPs have suboptimal policies rooted at s_0 that visit significantly fewer states than optimal ones. For example, many people often face the decision of choosing an airline ticket to a faraway destination. They typically have the option of flying there directly or with several stopovers. Flights with one or two stopovers are often cheaper than direct ones and therefore constitute a cost-minimizing plan. However, they also carry an increased probability of complications, e.g., missed connections, the necessity to get a transit visa, etc. Thus, overall one may expect a policy involving a multi-leg trip to have a higher probability of visiting many more states than a simpler but costlier one — taking a direct flight.

A priori, there is little reason to believe that an arbitrary inadmissible heuristic will help us find a smaller-than-optimal policy or, in general, cause heuristic search algorithms to visit fewer states than if it was guided by an admissible one. However, the inadmissible heuristics we describe below do appear to reduce the memory footprint of these algorithms. Like the admissible heuristics covered in Section 4.6, they are based on estimating the cost of plans in a determinized version of an MDP and, in turn, using these estimates to evaluate states of the MDP itself. However, the inadmissible heuristics we examine tend to make less crude assumptions when assessing the deterministic plan costs. On the one hand, this sometimes forces them to overestimate the value of states in the MDP and hence is the ultimate cause of their inadmissibility. On the other hand, it makes them more informative overall, i.e., makes them better able to tell advantageous states from inferior ones. Heuristic search by algorithms such as ILAO* and LRTDP translates this characteristic into fewer states visited while looking for the final policy, since it avoids states that seem subpar.

Unfortunately, it is hard to tell in advance whether the reduction in memory requirements due to a particular choice of algorithm and heuristic on a given MDP will be sufficient to avoid running out of space. Such a reduction is not certain even for online methods, and is even harder to predict for heuristic search. Nonetheless, as available memory permits, heuristic search with determinization-based inadmissible heuristics is a viable alternative to online algorithms for finding high-quality suboptimal policies.

6.3.1 h_{add}

Recall the delete relaxation from Section 4.6.2. This relaxation can be derived from an MDP by first determinizing it, i.e., by turning each probabilistic outcome of each of the MDP's actions into a separate deterministic action, and then removing all the negative literals from the actions' effects. Thus, literals in the delete relaxation can be only achieved, never "unachieved." We denote the resulting set of actions of the delete relaxation as $\mathcal{A}^r$. In Section 4.6.3, we saw an admissible heuristic based on the delete relaxation, h_{max}. The h_{max} heuristic effectively assumes that in the delete relaxation, the goal, a conjunction of literals $G_i, \ldots, G_n$, is only as hard to achieve as the "hardest" literal in it. This assumption makes the resulting state value estimates overly optimistic, since achieving the goal is typically costlier even in the delete relaxation, let alone the original MDP.

The first inadmissible heuristic we examine, h_{add} [30], assumes the other extreme — that achieving each goal literal in the delete relaxation does not in any way help achieve the others. Under this assumption, the goal literals are independent, and the cost of achieving the goal is the sum of costs of achieving each of the goal literals separately. Like h_{max}, h_{add} is applicable to MDPs where the cost of an action does not depend on the state in which the action is used, and thus can be captured by a single value $Cost(a)$. Therefore, h_{add}'s state value estimates are defined as

$$h_{add}(s) = \sum_{i=1}^{n} C(s, G_i), \tag{6.3}$$

where $C(s, G_i)$, the cost of achieving a goal literal when starting in state s, is defined recursively in terms of the costs of achieving other literals as

$$C(s, G_i) = \begin{cases} 0 & \text{if } G_i \text{ holds in } s, \\ \min_{a \in \mathcal{A}^r \text{ s.t. } G_i \text{ is in } add(a)} Cost(a) + C(s, prec(a)) & \text{if } G_i \text{ does not hold in } s \end{cases} \tag{6.4}$$

where

$$C(s, prec(a)) = \sum_{L_i \text{ s.t. } L_i \text{ holds in } prec(a)} C(s, L_i) \tag{6.5}$$

Notice that mathematically, h_{add} differs from h_{max} only by replacing maximization with summation in Equations 4.1 and 4.3. Thus, h_{add} can be computed in almost the same way and with the same efficiency. However, it may overestimate the cost of achieving the goal in the determinization and hence in the original MDP. Thus, it is inadmissible. Nonetheless, it tends to be more informative than h_{max} [30] and is therefore a good choice when solution quality guarantees can be sacrificed for scalability.

6.3.2 h_{FF}

The final heuristic derived from the delete relaxation that we consider in this book, and one of the most informative of this kind in practice, is h_{FF}. This heuristic is at the core of the FF planner [112], whose speed made practical the use of deterministic solvers in probabilistic planning. To produce a value estimate for a state s, h_{FF} assumes that at each time step, several actions in the delete relaxation can be applied concurrently. It also assumes that the MDP is given in a PPDDL-style factored form, and each action has a precondition limiting its applicability to a subset of the state space. Starting from s, h_{FF} iteratively applies all actions applicable in the current state, causing a "transition" to another state. This process yields a sequence of state transitions $s \to s_1 \to \ldots \to s_t \to s_g$, where s_g is the first state in the sequence containing all of the goal literals G_i.

After computing this sequence, h_{FF} estimates the total cost of all actions in it whose application was *necessary* at some point in order to achieve all goal literals from s. It does this by first considering the penultimate state in the sequence, s_t, going one-by-one through all actions applicable in it, and selecting those that achieve any of the goal literals in s_g that have not been achieved by the actions already selected during this step. This yields a set of actions A_t, minimal in the sense that each action in A_t is necessary to achieve at least one goal literal that can possibly be achieved directly from s_t.

At the next step, h_{FF} examines the preconditions of actions in A_t and computes a minimal set of actions A_{t-1} applicable in state s_{t-1} in the delete relaxation that achieves as many as possible of the literals in the goal *or* in the preconditions of actions in A_t. This procedure is repeated recursively to form a sequence of action sets $A_t, \ldots, A_0$, with the set A_0 consisting of actions applicable in s. By construction, actions in these sets are necessary to eventually achieve the goal, and h_{FF} lets $h_{FF}(s) = \sum_{j=0}^{t} \sum_{a \in A_j} C'(s_j, a)$, where $C'(s_j, a)$ gives the cost of applying a in s_j in the delete relaxation. $C'(s_j, a)$ can be defined in different ways. However, as in the cases of h_{max} and h_{add}, h_{FF} is mostly used for MDPs in which the cost of an action does not depend on the state where the action is applied and can be captured by a single value, $Cost(a)$. For these MDPs, we can write

$$h_{FF}(s) = \sum_{j=0}^{t} \sum_{a \in A_j} Cost(a). \tag{6.6}$$

The h_{FF} heuristic is not admissible, because the set of actions $\cup_{j=1}^{t} A_j$ is not necessarily a minimum-cost set of actions that achieves the goal literals. Finding such a set of the minimum cost is an NP-hard problem [51]. At the same time, h_{FF} is very informative, and under its guidance solvers typically explore many fewer states than with admissible heuristics.

6.3.3 h_{GOTH}

Our discussion of MDP heuristics started with h_{aodet} (Section 4.6.2). We claimed that h_{aodet} is informative but expensive to compute because it attempts to run a classical planner in order to solve the all-outcome determinization without relaxing it. Indeed, a heuristic may need to evaluate billions of states, and for each of them h_{aodet} would invoke a full-fledged classical planner. Any gains in MDP solver speed due to h_{aodet}'s informativeness would be offset by the enormous cost of calculating the heuristic itself.

Algorithm 6.3: Goal Regression

1 // $det\,Plan$ is the plan to be regressed
2 $det\,Plan \leftarrow [\langle prec_1, \langle add_1, del_1 \rangle \rangle, \ldots, \langle prec_m, \langle add_m, del_m \rangle \rangle]$
3 // Start regression with the goal conjunction
4 $r_{m+1} \leftarrow G_1 \wedge \ldots \wedge G_n$
5 **for** $i = m$ through 1 **do**
6 $\quad r_i \leftarrow \bigwedge [(set(r_{i+1}) \setminus (set(add_i) \cup set(del_i))) \cup set(prec_i)]$
7 **end**
8 return $r_1, \ldots, r_m$

The situation would be different if we could efficiently use each invocation of a classical planner to evaluate many states, not just one, as h_{aodet} does. This would be equivalent to amortizing the cost of calling the planner. If, on average, one classical planner call can help evaluate sufficiently many states, employing this strategy may well pay off due to the informativeness of its state estimates.

This idea of *generalizing* value estimates across many states is at the heart of the GOTH heuristic, h_{GOTH} [132; 134]. Like all other determinization-based heuristics we discussed, h_{GOTH} assumes that the MDP we are trying to solve is in a PPDDL-style factored form, i.e., its actions have preconditions dictating in which states these actions are applicable. Recall that in h_{aodet}, the value estimate for state s in MDP M is the cost of a plan from s to the goal in the all-outcome determinization M_d^a. The main insight of h_{GOTH} is that the same plan reaches the goal not only from s but from many other states as well. To see why this is so, observe that each such plan can be viewed as a "super-action," a composition of several action applications. Like an ordinary action, it has an effect (the goal literal conjunction) and a precondition that generally holds in several states. If we compute this precondition for a goal plan, we can store it in a table with the cost of the corresponding plan. When we need to evaluate a state s, we can check the preconditions from this table to see if any of them holds in s, and produce a heuristic value for s by aggregating the costs associated with all such preconditions that hold in s.

This is exactly what h_{GOTH} does. It maintains a table of goal plan preconditions and their costs. When asked to evaluate a state s, h_{GOTH} first checks whether any of the stored plan preconditions hold in it. If so, it takes the minimum of costs associated with these plan preconditions and returns this value as $h_{GOTH}(s)$. If not, it launches a classical planner from s, computes its precondition, stores it in the table with the plan's cost, and returns the plan's cost. As $h_{GOTH}(s)$'s library of plan preconditions and costs grows, it is able to evaluate more and more states without resorting to a classical planner, i.e., very quickly.

To compute a goal plan precondition, h_{GOTH} resorts to *regressing the goal through the plan*. Before presenting the goal regression algorithm, we introduce some notation. We will view a goal plan in M_d^a as a sequence $[\langle prec_1, \langle add_1, del_1 \rangle \rangle, \ldots, \langle prec_m, \langle add_m, del_m \rangle \rangle]$, where $prec_i$ is the precondition of the i-th action of the plan, and add_i, del_i is the effect of that action. Further, we let $set(C)$ denote the set of literals in some literal conjunction C. Finally, we let $\bigwedge \mathcal{K}$ stand for the conjunction of all literals contained in some set of literals $\mathcal{K}$.

Goal regression (Algorithm 6.3) finds the preconditions for all suffixes of the given goal plan. That is, assuming the given plan is of length m, goal regression discovers the precondition r_i for each subplan of the form $[\langle prec_i, \langle add_i, del_i \rangle \rangle, \ldots, \langle prec_m, \langle add_m, del_m \rangle \rangle]$ for i ranging from 1 to m. In particular, it discovers the precondition r_1 for the whole plan itself, which is the plan's own suffix for $i = 1$. Goal regression starts by taking the goal conjunction (r_{m+1} in Algorithm 6.3), removing from it the effects of the last action in the plan, and conjoining the action's precondition to the result (line 6). What we get is another literal conjunction, which can be viewed as the "new goal conjunction." Applying this operation iteratively until the first action of the plan (lines 5–7) yields the desired sequence of plan suffice preconditions $r_1, \ldots, r_m$. The cost associated with each r_i is the cost of the plan suffix starting at i.

Generalizing plan costs via plan preconditions allows h_{GOTH} to provide informative state values at a relatively low computational price. For added efficiency, h_{GOTH} uses fast (but suboptimal) classical planners such as FF to compute the goal plans in the determinization. The suboptimality of these planners, along with the use of generalization, make h_{GOTH} an inadmissible heuristic.

6.4 DIMENSIONALITY REDUCTION-BASED TECHNIQUES

The amount of reduction in space requirements due to the use of an inadmissible heuristic strongly depends on the properties of the particular MDP at hand. If all suboptimal policies rooted at s_0 in this MDP visit a comparable or larger number of states than the optimal ones, no heuristic will be very helpful in reducing the memory footprint of the solver. This motivates us to consider a yet another class of approximation techniques, those based on dimensionality reduction. As already briefly mentioned, the central idea of dimensionality reduction is to represent an MDP's policy/value function with a small number of parameters, much smaller than the number of states this policy may visit.

In particular, recall from Section 2.5.4 that, since our MDP is factored, we can represent its value functions in a factored form as well. Factorization lets us write down the value functions in a compact, possibly parametrized way. We already saw one example of such a factored value function in Section 2.5.4, $V^*((X_1, X_2)) = X_1 + X_2$, which represents a state's value as a sum of the state variable values. This value function is an instance of a family of linear value functions. They have the form $V^*((X_1, \ldots, X_n)) = \sum_{i=1}^{n} w_i X_i$, where $\{w_i\}_{i=1}^{n}$ are parameters. In this case, the number of parameters, n, is linear in the number of state variables, while the number of parameters of a value function in a flat representation can be as high as the size of the state space, 2^n. Thus, a factored representation may yield an exponentially large reduction in the dimensionality of the value function, and hence in space requirements. Generally, the factored value function can be much more complicated than linear, and the precise number of parameters may vary depending on the problem. However, the parametrization of the value function is typically picked so that the parameters are easy

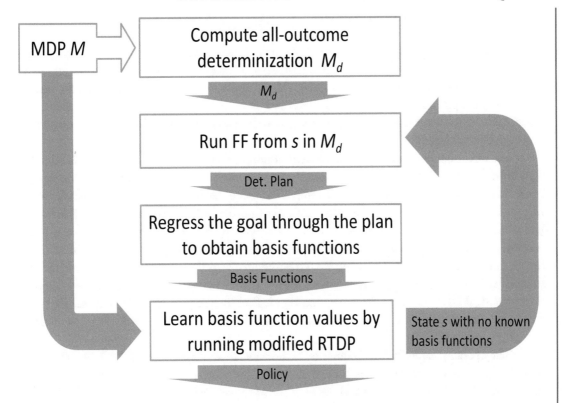

Figure 6.2: ReTrASE

to learn, and so that their number increases very slowly with problem size. A policy can be factored in a similar way.

In this section, we will examine several algorithms that are instances of dimensionality reduction. In addition, we will reinterpret the h_{GOTH} heuristic, an algorithm that we have already discussed in a different context, as dimensionality reduction as well.

6.4.1 RETRASE

Besides domain determinization, the h_{GOTH} heuristic we saw at the end of the previous section implicitly exploits another common paradigm, whose usefulness extends far beyond generating state initialization values. This paradigm is called *function approximation*, and is itself a flavor of dimensionality reduction. Its main strategy is to approximate the possibly very complicated value function as a combination of several simpler functions that also assign values to states. These "building blocks" are called *basis functions*. Notice that each plan precondition r computed by h_{GOTH} defines a basis function b_r. This basis function has the value of 1 in all states where the corresponding plan precondition holds, and ∞ in all other states. With the help of the basis functions, h_{GOTH} defines the value function $h_{GOTH}(s) = \min_{r \in \mathcal{U}} w_r b_r(s)$, where $\mathcal{U}$ is the set of all plan preconditions that h_{GOTH} has discovered, and w_r is the cost of the plan whose precondition is r. That is, to aggregate the basis functions into a heuristic value function, h_{GOTH} uses the min operator, and the parameters of the value function are the basis function weights. Thus, as long as the number of discovered basis functions (plan preconditions) is smaller than the number of states, function approximation saves space. Clearly, the same principle can be used to write down not only the initial, heuristic value function, but any value function in general.

We next examine several planners based on function approximation. The first of them is an algorithm called ReTrASE [131; 134]. Fundamentally, ReTrASE is very similar to the h_{GOTH} heuristic. Like the latter, it computes and stores a number of basis functions derived from deterministic goal plan preconditions, which capture causally important information about the problem. Each basis function has a weight associated with it. The weights let ReTrASE derive state value the same way h_{GOTH} does — by taking the minimum over the weights of those basis functions that hold in the given state. The difference between ReTrASE and h_{GOTH} lies in how they determine the basis function weights. Whereas h_{GOTH} estimates the weight of a basis function once and never modifies it afterward, ReTrASE *learns* these weights via an RTDP-like procedure. In a sense, ReTrASE combines several paradigms in probabilistic planning — dimensionality reduction, determinization, and heuristic search.

The operation of ReTrASE is described schematically in Figure 6.2. ReTrASE's central component is a modified version of the RTDP algorithm (Section 4.4.1). It runs a series of trials the same way as RTDP, but performs updates over basis function weights [131] instead of state values and uses basis function weights instead of a state-value table to evaluate states. When ReTrASE encounters a state s for which it does not have basis functions, it uses a classical planner, e.g., FF, to look for a goal plan from s in the all-outcome determinization M_d^a (Section 4.6.1) of the MDP. Then it regresses the goal through the resulting plan (if one is found) using Algorithm 6.3, and stores the basis functions derived in this way. At least one of these basis functions is guaranteed to have value 1 in s, giving ReTrASE a way to evaluate this state. Once ReTrASE discovers a new basis function, it starts performing modified Bellman backups on it as part of the RTDP trials. When stopped, ReTrASE can return a policy by reading it off of the basis-function-induced value function.

Thus, as in the case of h_{GOTH}, the parameters ReTrASE relies on are the basis function weights. Their number is not fixed in advance; ReTrASE keeps discovering new basis functions as long as it has memory and time. ReTrASE's key insight is that the number of "important" basis functions, those that ReTrASE is likely to discover, is typically much smaller than the number of states. (Theoretically, however, the number of *possible* basis functions derived from plan preconditions in a binary factored MDP is $3^{|\mathcal{X}|}$, where $\mathcal{X}$ is the number of state variables, i.e., is larger than the number of states.) As a consequence, the amount of information ReTrASE has to store is usually much smaller than for algorithms that represent the value function as an explicit state-value table.

6.4.2   APPROXIMATE POLICY ITERATION AND LINEAR PROGRAMMING

Both h_{GOTH} and ReTrASE are built around a very specific kind of basis functions — those that are derived via goal regression. In goal-oriented scenarios, these basis functions often make sense because they capture goal reachability information particular to the problem at hand. However, there are plenty of settings where basis functions of this kind are not so informative. For instance, consider again the problem of managing n servers, Sysadmin (Section 2.5.3). Suppose the servers are connected in a network with a certain topology (each server is connected to several others). At each time step, each server may fail with some probability. A server's shutdown increases load on its neighbors in the network, raising their probability of failure. The job of the agent (the network administrator) is to maintain as many servers running simultaneously as possible by fixing and restarting servers that fail. The agent can repair only so many servers per unit time, so it invariably needs to decide which of the failed servers are affecting the network the most and fix them first, making the problem nontrivial. The objective of maintaining the network in an operational condition is not a goal-oriented one. It is more appropriately modeled with an infinite-horizon discounted-reward MDP (Section 2.3.3) where the agent gets a reward at each time step, proportional to the number of servers currently running and tries to maximize a discounted expected sum of these rewards. Although any discounted MDP can be turned into an SSP problem (Theorem 2.21) by adding an artificial goal state, this goal has no semantic meaning, and the regression-derived basis functions would not make much sense here. ReTrASE would not do well on such a problem.

In domains such as this one, it is helpful to have basis functions designed by a human expert aware of the semantics of the problem. E.g., in Sysadmin, the quality of a state is correlated with the number of running servers. Indeed, barring pathological cases, it is easier to support a network with most machines already running than with most machines down. Therefore, the more machines are up in a state, the more long-term rewards one generally can collect starting from it. Accordingly, one meaningful set of basis functions would be $\mathcal{B} = \{b_i | 1 \leq i \leq n\}$, where

$$b_i(s) = \begin{cases} 1 & \text{if server } i \text{ is running in } s, \\ 0 & \text{otherwise} \end{cases} \tag{6.7}$$

For this set of basis functions, we could define an optimal value function approximation to V^* to be

$$\tilde{V}^*(s) = \sum_{i=1}^{n} w_i b_i(s) \tag{6.8}$$

for some weights $w_i, \ldots, w_n$ chosen so that $\tilde{V}^*$ closely approximates V^*. In order to find $\tilde{V}^*$, we need to answer a general question: how do we find weights for an arbitrary set of basis functions $\mathcal{B}$ that result in a good approximation?

Two algorithms for doing this are Approximate Policy Iteration and Approximate Linear Programming (API and ALP respectively). There are several flavors of each, differing primarily in how they measure the quality of an approximate value function. We consider the versions of API and ALP that assess an approximation $\tilde{V}$ w.r.t. a target value function V in terms of the $\mathcal{L}_\infty$-*norm*, defined as $||V - \tilde{V}||_\infty = \max_{s \in \mathcal{S}} |V(s) - \tilde{V}(s)|$.

Approximate Policy Iteration. API [102] closely resembles its exact counterpart, PI (Section 3.3). As a reminder, PI repeatedly executes a sequence consisting of two steps, *policy evaluation* and *policy improvement* (Algorithm 3.2). During policy evaluation, PI computes the value function V^{π_j} corresponding to the policy π_j produced in the current, j-th iteration. During policy improvement, PI constructs a policy π_{j+1} greedy w.r.t. V^{π_j}. API also iterates over the policy evaluation and policy improvement steps. Moreover, its policy improvement step is identical to PI's.

What API does differently is policy evaluation. Remember, our objective is to compute a good approximation $\tilde{V}^*$ to V^* (in the $\mathcal{L}_\infty$-norm) that can be represented as a linear combination of the basis functions from a set $\mathcal{B}$ specified by an expert. To find $\tilde{V}^*(s)$, at each policy evaluation step API computes $\tilde{V}^{\pi_j} = \sum_{i=1}^{n} w_i^{(j)} b_i$ that best represents the value function of the current policy, V^{π_j}. API's policy improvement step then computes π^{j+1} greedy w.r.t. $\tilde{V}^{\pi_j}$, and the process continues. Thus, API generates a sequence of value function approximations $\{\tilde{V}^{\pi_j}\}_{j=1}^{\infty}$ in the hope that the sequence converges to a reasonable $\tilde{V}^*$.

The key part of API is finding a good approximation $\tilde{V}^{\pi_j}$ over the given set of basis functions. API's approach to doing this is analogous to PI's policy evaluation in the following sense. If we extend the ideas of Section 3.2.1 to discounted-reward MDPs, exact policy evaluation for such an MDP amounts to solving a system of linear equations

$$V^\pi(s) - \sum_{s' \in \mathcal{S}} \mathcal{T}(s, \pi(s), s') \left[\mathcal{R}(s, \pi(s), s') + \gamma V^\pi(s') \right] = 0 \text{ for all } s \in \mathcal{S} \qquad (6.9)$$

Hence, $V = V^\pi$ can be thought of as the value function that achieves the minimum of the difference $|V(s) - \sum_{s' \in \mathcal{S}} \mathcal{T}(s, \pi(s), s') \left[\mathcal{R}(s, \pi(s), s') + \gamma V(s') \right]|$ for every state, which happens to be 0. Mathematically,

$$V^\pi = \underset{V}{\operatorname{argmin}} \left[\max_{s \in \mathcal{S}} \left| V(s) - \sum_{s' \in \mathcal{S}} \mathcal{T}(s, \pi(s), s') \left[\mathcal{R}(s, \pi(s), s') + \gamma V(s') \right] \right| \right] \qquad (6.10)$$

By analogy, to find a good approximation $\tilde{V}^{\pi_j} = \sum_{i=1}^{n} w_i^{(j)} b_i$, API finds a weight vector $\vec{w} = (w_1, \ldots, w_n)$ satisfying

$$\vec{w}^\pi = \underset{\vec{w}}{\operatorname{argmin}} \left[\max_{s \in \mathcal{S}} \left| \sum_{i=1}^{n} w_i b_i(s) - \sum_{s' \in \mathcal{S}} \mathcal{T}(s, \pi(s), s') \left[\mathcal{R}(s, \pi(s), s') + \gamma \sum_{i=1}^{n} w_i b_i(s') \right] \right| \right] \quad (6.11)$$

As it turns out, the minimization in Equation 6.11 can be cast as a linear program in n variables $(w_1, \ldots, w_n)$ with $2|\mathcal{S}|$ constraints [102]. To do policy evaluation in the j-th iteration, API solves this linear program and finds the corresponding weights $\vec{w}^{\pi_j}$.

Unlike PI, theoretically API is not guaranteed to converge. It may forever oscillate between several policies, because policies π_j and $\pi_{j'}$ produced at iterations j and $j' > j$ may have $\tilde{V}^{\pi_j} = \tilde{V}^{\pi_{j'}}$ for the given basis function set $\mathcal{B}$. Thus, the j'-th iteration of the algorithm may effectively reset API to where it was after the j-th iteration.

Nonetheless, if API does converge, as is often the case in practice, we can bound the error of the resulting policy. Besides producing a weight vector $\vec{w}^{\pi_j}$, solving the aforementioned linear program during each policy evaluation step lets us calculate a quantity $\phi_j = \max_{s \in \mathcal{S}} | \sum_{i=1}^{n} w_i^{\pi_j} b_i(s) - \sum_{s' \in \mathcal{S}} \mathcal{T}(s, \pi(s), s') \left[\mathcal{R}(s, \pi(s), s') + \gamma \sum_{i=1}^{n} w_i^{\pi_j} b_i(s') \right] |$, which is the *approximation error* under the $\mathcal{L}_\infty$-norm at the j-th iteration of API. Assuming API halts after K iterations because its policy does not change anymore, the following result holds for the value function $\tilde{V}^*$ of policy $\tilde{\pi}^*$ produced by API [102]:

$$||V^* - \tilde{V}^*||_\infty \leq \frac{2\gamma (\max_{1 \leq j \leq K} \phi_j)}{1 - \gamma} \tag{6.12}$$

Approximate Linear Programming. We can derive ALP from LP in the same manner as we derived API from PI, by substituting $\tilde{V}^* = \sum_{i=1}^{n} w_i b_i$ for V^* into the original LP formulation from Section 3.7 adapted for discounted-reward MDPs. The resulting linear program looks as follows:

Variables	$w_i \quad \forall i, 1 \leq i \leq n$
Minimize	$\sum_{s \in \mathcal{S}} \alpha(s) \sum_{i=1}^{n} w_i b_i(s)$
Constraints	$\sum_{i=1}^{n} w_i b_i(s) \geq \sum_{s' \in \mathcal{S}} \left[\mathcal{R}(s, a, s') + \gamma \mathcal{T}(s, a, s') \left[\sum_{i=1}^{n} w_i b_i(s) \right] \right] \quad \forall s \in \mathcal{S}, a \in \mathcal{A}$

Again, notice that the number of variables is now only $n = |\mathcal{B}|$, not $|\mathcal{S}|$. However, as with API, this achievement comes at a cost. In the original LP formulation, the values $\alpha(s)$ did not matter as long as they were positive — LP would yield the optimal value function for any setting of them. Contrariwise, ALP heavily depends on their values, as they determine the approximation quality. Moreover, there is no known general way of setting them in a way that would provably result in good approximation weights $\vec{w}$ [102].

Making API and ALP More Efficient. The motivation for all the approximation algorithms we have considered is to be able to solve larger MDPs within available time and memory constraints. In particular, the dimensionality reduction techniques are supposed to reduce memory requirements by representing the solution policy with a small number of parameters and using an amount of space polynomial in that number. With this in mind, let us take a closer look at API and ALP. Both algorithms ultimately amount to solving a linear program (in the case of API — multiple times, during the policy evaluation steps). The number of variables in these LPs is $n = |\mathcal{B}|$, the size of the basis function set $\mathcal{B}$. Thus, if the number of basis functions is small, the policy representation is fairly compact, as we intended. However, consider the number of constraints in these LPs. For API and ALP alike, the LPs they have to solve involve at least one constraint *for every state*; despite the representational compactness, these algorithms still scale as poorly as VI and PI!

Fortunately, with additional restrictions on the basis function set $\mathcal{B}$ we can make API and ALP much more efficient. These restrictions critically rely on the fact that we are dealing with *factored* MDPs, i.e., each state is a conjunction/vector over a set of variables $\mathcal{X}$. So far, we have implicitly assumed that each basis function $b \in \mathcal{B}$ needs the values of all $|\mathcal{X}|$ variables to produce a value for the state. In the meantime, for many problems, meaningful basis functions are very "local," i.e., they rely on the values of only a few of the state variables each. The Sysadmin scenario serves as a prime example. For this domain, each basis function b_i in the set $\mathcal{B}$ we have come up with needs the value of only *one* state variable in order to determine the state value — the value of the variable reflecting the status of the i-th server in the network.

In general, let us define the *scope* of a basis function b as the set $\mathcal{X}_b \subseteq \mathcal{X}$ of state variables whose values b needs in order to produce the value of a state. Further, let us restrict each basis function to be local, i.e., to have $|\mathcal{X}_b| < l$ for some $l < |\mathcal{X}|$. The central scalability issue of the API and ALP versions presented earlier is caused by the number of their LP constraints being exponential in the size of the scope of their basis functions, so far assumed to be $|\mathcal{X}|$. A significant improvement to these algorithms lies in the nontrivial observation [102] that if the largest basis function scope is of size l, the LPs in API and ALP can be replaced by alternative formulations with the number of constraints exponential in l, not $|\mathcal{X}|$. Constructing these alternative LPs is an involved process, and we refer the reader to the original paper [102] for more information. However, the effect of this modification can be appreciated even without knowing its details. For instance, in the Sysadmin domain it makes an exponential difference in required memory compared to the original API and ALP.

6.4.3   FPG

FPG [50] belongs to a class of algorithms that learn a policy directly (like PI), not via a value function (like VI and heuristic search techniques). It tries to construct a *probabilistic Markovian policy* $\pi : \mathcal{S} \times \mathcal{A} \rightarrow [0, 1]$, one that gives a probability distribution over actions for each state. Recall that an optimal policy for an SSP MDP does not actually need to be that complicated — it only needs to specify a single action to execute in each state (Theorem 2.18). However, as the example of the FPG algorithm shows, policies of the form $\pi : \mathcal{S} \times \mathcal{A} \rightarrow [0, 1]$ may be easier to learn.

Algorithm 6.4: FPG

1 $s \leftarrow s_0$
2 // Initialize the gradient vector
3 $\vec{e} \leftarrow [0]$
4 **for** all $a \in \mathcal{A}$ **do**
5 | Initialize $\vec{\theta}_a$ randomly
6 **end**
7 $\vec{\theta} = concatenate_{a \in \mathcal{A}}\{\vec{\theta}_a\}$
8 **while** there is time left and $\pi_{\vec{\theta}}$ has not converged **do**
9 | $a \leftarrow$ sample an action according to distribution $\pi_{\vec{\theta}}(s, a)$
10 | $s' \leftarrow$ execute action a in s
11 | // Compute a gradient approximation
12 | $\vec{e} \leftarrow \beta\vec{e} + \nabla \log \pi_{\vec{\theta}}(s, a)$
13 | // Update the parameters of the policy
14 | $\vec{\theta} \leftarrow \vec{\theta} + \alpha\mathcal{C}(s, a, s')\vec{e}$
15 | **if** $s' \in \mathcal{G}$ **then**
16 | | $s \leftarrow s_0$
17 | **end**
18 | **else**
19 | | $s \leftarrow s'$
20 | **end**
21 **end**
22 return $\pi_{\vec{\theta}}$

To compactly specify a policy $\pi : \mathcal{S} \times \mathcal{A} \rightarrow [0, 1]$ for a factored MDP, FPG associates a parameterized *desirability function* $f_{a|\vec{\theta}_a} : \mathcal{S} \rightarrow \mathbb{R}$ with each action a. This function describes how "good" a is in each state. Given such a function for each action, FPG defines

$$\pi_{\vec{\theta}}(s, a) = \frac{e^{f_{a|\vec{\theta}_a}(s)}}{\sum_{a' \in \mathcal{A}} e^{f_{a'|\vec{\theta}_{a'}}(s)}}, \qquad (6.13)$$

where the vector $\vec{\theta}$ that parameterizes $\pi_{\vec{\theta}}(s, a)$ is a concatenation of all actions' parameter vectors $\vec{\theta}_a$. Intuitively, $\pi_{\vec{\theta}}(s, a)$ is a softmax distribution based on the "desirability" of various actions in a given state. The higher an action's desirability in s, the more likely it is to be chosen in s for execution. When we are dealing with a PPDDL-style factored MDP, whose actions have preconditions limiting their applicability to certain states, $\pi_{\vec{\theta}}(s, a) = 0$ whenever a is not applicable in s, and the summation in the denominator in Equation 6.13 is taken only over actions applicable in s.

Thus, the key aspect of FPG is the representation and learning of the desirability functions. The dependence between the inputs of these functions (i.e., vectors of variable values representing states) and their outputs (real numbers) can have virtually any form. Consequently, FPG stores these functions as neural networks, a model with a large representational power. The topology of the networks is the same for all actions, but the node weights vary from action to action and are contained in the parameter vector $\vec{\theta}_a$. The space savings of FPG compared to VI-based approaches depend on the size of $\vec{\theta}_a$ for each a. For factored MDPs, the length of $\vec{\theta}_a$ is proportional to the number of state variables $|\mathcal{X}|$. This means that whereas VI-based methods need $O(|\mathcal{S}|) = O(2^{|\mathcal{X}|})$ space to store a policy for them, FPG only needs $O(|\mathcal{X}|) = O(\log_2(|\mathcal{S}|))$ space — a very large reduction in memory requirements.

To learn the parameter vectors, FPG (Algorithm 6.4) uses gradient descent, a standard approach for learning neural networks. At a high level, each vector $\vec{\theta}$ induces a policy $\pi_{\vec{\theta}}$ with an associated value at the initial state, $V_{\vec{\theta}}^{\pi}(s_0)$. FPG looks for $\vec{\theta}$ whose corresponding policy has the optimal $V_{\vec{\theta}}^*(s_0)$. To do so, FPG starts with $\vec{\theta}$ that induces a uniformly random policy, assigning an equal probability to all (applicable) actions in any state. The algorithm then iteratively computes an approximation $\vec{e}$ of the gradient for the current policy with respect to the parameters $\vec{\theta}$ of the desirability functions [15; 50] and uses it to update the parameter vector $\vec{\theta}$, thereby altering the policy. To implement this iterative process, FPG simulates the current policy the same way algorithms such as LRTDP do, by running a series of trials (lines 8-21 of Algorithm 6.4). During each trial, the actions recommended by the changing policy (i.e., ultimately, by the desirability functions), incur some cost. After each action execution, FPG uses the incurred cost sequence incurred so far to compute a gradient approximation $\vec{e}$ (line 12). This approximate gradient (computing the actual gradient would be very expensive) takes into account both the gradient's previous value and the change in the gradient due to the latest action execution ($\nabla \log \pi_{\vec{\theta}}(s, a)$). The weight of the old gradient value in this combination is determined by the tunable β parameter — the bigger it is, the slower FPG converges but the fewer erratic policy adjustments it makes. Intuitively, the gradient approximation suggests modifying $\vec{\theta}$ to deter high-cost actions and encourage low-cost ones, so that changing $\vec{\theta}$ along the gradient hopefully causes the desirability functions to recommend a better policy. Modifying $\vec{\theta}$ accordingly (line 14) gives FPG a new policy, which recommends the next action to execute, and the gradient update process repeats.

While the original FPG version [48] chooses actions recommended by the current policy, as above, another variant called FF+FPG [49] uses the FF planner to guide the exploration. FF plays the role of a teacher, showing FPG what sequences of actions can reach the goal when FPG's own initial nearly random policies have difficulty getting to the goal states. The use of FF helps FF+FPG converge faster by focusing the learning procedure on promising actions.

6.5   HIERARCHICAL PLANNING

We now discuss a popular set of MDP techniques that leverage a hierarchical decomposition of a large problem into smaller subproblems. The key intuition behind this set of techniques comes from the way humans solve planning problems.

What are the characteristics of the way humans plan? How do we solve a new problem? We of course never solve a full blown MDP in one go. In a complex environment, we often reduce a problem into smaller subproblems, solutions of which can be combined to find a complete solution.

In fact, this subdivision can be at multiple hierarchical levels. For example, in a simple navigation problem, we may first decide that the overall plan will involve the following high-level actions or subgoals: going to the car, entering the car, driving to the destination, parking the car at a nearby spot, and finally navigating from the car to the destination. These high-level actions may be further divided into more fine-grained actions and so on. For instance, entering the car may further subdivide into finding the key, opening the door, sitting in the car, and closing the door.

Another important feature of our decision making is an understanding of the relevant parts of a state for a specific decision problem. For example, if our goal is to walk to the car, we will realize that the location of the keys or the color of the car are irrelevant. Thus, not only do we decompose a problem into subproblems, but we also abstract away each subproblem so that it contains only the relevant state variables. In other words, we naturally perform *state abstraction* within the hierarchical *problem decomposition*.

A natural way to design approximate solvers is to mimic this kind of human decision making. To operationalize it, we need to answer several important questions. What is the best way to represent a problem decomposition? How can we leverage a known decomposition for fast planning? How do we characterize the qualities of such solutions? Can we *automatically* learn a problem decomposition (and state abstraction), or does it have to be provided by a domain designer? Addressing them is the focus of this section.

Most hierarchical approaches were developed in the context of infinite-horizon discounted-reward maximization MDPs, so we use the framework from Section 3.8 for this exposition. Moreover, many of the approaches were originally developed in the reinforcement learning literature [13]. Since the issues for hierarchical planning are similar in AI planning and reinforcement learning literatures, in the following discussion we take the planning view. An interested reader may consult the survey on hierarchical reinforcement learning techniques [13] for further details.

6.5.1   OPTIONS

One way for a domain designer to provide domain knowledge is to encode complete control routines for important behaviors. Imagine a robot that has predefined modules for object manipulation, entertainment (like a dance sequence to please the audience) and so on. This kind of knowledge is best encoded in an *option* representation [225].

Definition 6.7   **Option.** An *option* o is a tuple $\langle \mathcal{I}_o, \pi_o, \beta_o \rangle$, where

- $\mathcal{I}_o \subset \mathcal{S}$ is the set of states in which this option may be invoked,

- $\pi_o : \mathcal{S} \to \mathcal{A}$ is the policy executed by an agent within this option,[2]

- $\beta_o : \mathcal{S} \to [0, 1]$ is the termination condition; in particular, $\beta_o(s)$ represents the probability that o will terminate in state s.

---

[2]Researchers have also studied extensions where the policy is stochastic or when the policy additionally depends on the entire history [225].

An example of an option for our navigation domain is a 'get inside the car' routine. This option may be initiated ($\mathcal{I}_o$) when the agent is standing outside the car. The policy (π_o) would list the action to take in each state within this option. The option may terminate (β_o) deterministically when the agent is inside the car or stochastically after some time if the agent cannot find the keys.

An option-based controller comprises two levels (or, in general, multiple levels, if a higher-level option defines a policy over lower-level options). The higher-level controller chooses the best option available in the current state, after which the control passes to the controller of the invoked option. This lower-level controller executes the actions within the option. Eventually the option terminates, whereupon the control returns to the higher-level controller, which decides to choose another option based on the current state and so on. If all the options are completely pre-specified then planning happens only at the highest level.

Thus, the operation of a basic options-based planner can be summarized as follows:

Input: Such a planner takes in a discounted-horizon infinite-reward or an SSP MDP, a set of options for it, and a fully specified policy for each option.

Output: The output produced by an options-based planner is a policy that may either only use options, or both the options and the primitive actions specified as part of the MDP.

Planning: Since the option policy π_o is completely specified, for the purposes of an undiscounted SSP planner, an option can be thought of as just another action, with its known applicable states. Its termination outcomes are also known, and the outcome probabilities can be computed. The Bellman Equations for option-based planning are the same as those for SSP MDPs.

However, in the case of discounted-reward problems, the Bellman equations need an important change. Recall that for these problems each subsequent state value is discounted by γ. Now, since an option may take multiple steps to execute, we cannot just discount the values of the option's terminating states by γ; we need to additionally discount based on the number of steps an option executed before termination. Such problems in which an action may execute for multiple time-steps are modeled as a Semi-MDP [115; 197]. Thus, planning with options is technically a Semi-MDP. The Bellman equations get modified as:

$$Q^*(s, a) = \sum_{s' \in \mathcal{S}, N} \mathcal{T}_o(s, s', N) \left[R_o(s, s', N) + \gamma^N V^*(s') \right] \tag{6.14}$$

In this equation, N denotes the number of steps for which the option executes and $R_o(s, s', N)$ denotes the expected discounted reward collected within the option started at s that terminates at s' in N steps. $\mathcal{T}_o$ similarly denotes the probability that o initiated at s terminates at s' at the Nth step. A Semi-MDP VI algorithm (and extensions) is similar to its analogous MDP algorithm with this modified Bellman backup [42]. We can use these algorithms to obtain a policy over options. The main challenge is in computing the option models, R_o and $\mathcal{T}_o$.

We first optimize Equation 6.14 so that we can save on some computation. The main idea is to use the linearity of expectation (the expected value of the sum of two random variables is equal to the sum of individual expected values of the two variables) to equivalently write Equation 6.14 as:

$$Q^*(s, a) = \mathcal{R}_o(s) + \gamma \sum_{s' \in \mathcal{S}} \mathcal{T}_o^\gamma(s, s') V^*(s') \tag{6.15}$$

Here, $\mathcal{R}_o(s)$ represents the total expected discounted reward obtained with the option o when started in state s. This is easily defined as a set of recursive equations:

$$\mathcal{R}_o(s) = \sum_{s' \in \mathcal{S}} \left[\mathcal{R}(s, \pi_o(s), s') + \gamma(1 - \beta_o(s'))\mathcal{R}_o(s') \right] \tag{6.16}$$

That is, the total reward at state s is the immediate reward obtained by taking the action $\pi_o(s)$ plus the long-term expected discounted reward in case the option does not terminate at s', which happens with probability $1 - \beta_o(s')$.

Returning to Equation 6.15, $\mathcal{T}_o^\gamma(s, s')$ in it is a *discounted* transition model of terminating in s', governed by the following identity (adapted from [108]):

$$\mathcal{T}_o^\gamma(s, s') = \mathcal{T}(s, \pi_o(s), s')\beta_o(s') + \gamma \sum_{s'' \in \mathcal{S}} \mathcal{T}(s, \pi_o(s), s'')(1 - \beta_o(s''))\mathcal{T}_o^\gamma(s'', s') \tag{6.17}$$

An interested reader may verify that Equation 6.15 returns the same values as Equation 6.14. Notice that all the equations 6.15-6.17 are essentially linear equations and can be easily solved using policy evaluation techniques from Section 3.2. Thus, we can use VI-like methods for obtaining a plan over the set of options.

Solution Quality: An important question to ask is: can we get an optimal solution if we are planning with options? Since we impose an additional hierarchy on the existing MDP, in general the solution found is the best of all solutions consistent with the hierarchy. This solution may not be globally optimal. For instance, it is clear that if the input options are not that good, then a globally optimal solution may not even be expressible as a policy over options. Thus, it is useful to define a weaker notion of optimality known as *hierarchical optimality*. A hierarchically optimal policy is defined as the best policy consistent with the given hierarchy. An optimal Semi-MDP algorithm over the options computes a hierarchically optimal policy.

Researchers have also studied ways of improving a hierarchically optimal policy. One approach adds each primitive action as a one-step option to the set of options applicable in each state. Here, an optimal policy over combined options $(\mathcal{A} \cup \mathcal{O})$ is bound to compute the global optimal ($\mathcal{O}$ is the set of all options), since now all actions can be individually composed to construct the best policy. On the other hand, by using both the sets, the advantage of the options framework in reducing the branching factor is lost. In fact, the branching factor has necessarily increased compared to the flat problem. Still, options may provide valuable savings because they allow the agent to achieve a longer lookahead in one step. The tradeoff between solution quality and planning time is an empirical question and designers may or may not prefer to use the different sets depending upon the domain and available planning time.

A second method for improving the solution quality is to terminate an option early if another option seems more beneficial. This approach has its roots in an early work on hierarchical distance to goal discussed in Section 6.5.4. Here, the controllers may be modified so that they do not wait until the termination of an option. Instead, the high-level controller can choose to switch to another option. Mathematically, this is achieved by creating a new set of options $\mathcal{O}'$, which has an o' for each o. If the original policy over options $\mathcal{O}$ was π, we define a new policy π' over options $\mathcal{O}'$ so that $\pi'(s, o') = \pi(s, o')$ for all $s \in \mathcal{S}$. The option o' is similar to o except that it terminates whenever switching an option is more valuable, i.e., when $Q^\pi(s, o) < V^\pi(s)$ (for maximization MDPs). Technically for these states s, we set $\beta_{o'}(s) = 1$. Such a policy π' is known as an *interrupted policy* of π. It can be proved that in such a scenario, $V^{\pi'}(s) > V^\pi(s)$, i.e., the interrupted policy is a better policy. This gives us a computationally efficient mechanism to improve the quality of a hierarchically optimal policy.

An advantage of the options framework is that it allows a complete specification of behaviors or macro-actions. This can significantly reduce the search space of the planner and can be very useful in creating practical agents. For example, options have been used in robotic control problems requiring robots to transfer skills gained in one setting to another [138]. At the same time, the key drawback of the framework is the requirement of a complete specification for the options. This may put undue stress on a domain designer. To alleviate this, methods have been developed for learning option policies. We discuss these briefly in Section 6.5.6.

6.5.2   TASK HIERARCHY

Another popular representation for hierarchical planning is a *task hierarchy*, in which a domain designer decomposes the problem into a set of unordered lower-level *subtasks* [79].

Definition 6.8   Subtask. A *subtask t* is a tuple $\langle \mathcal{S}_t, \beta_t, \mathcal{R}_t \rangle$, where

- $\mathcal{S}_t \subset \mathcal{S}$ is a subset of the state space $\mathcal{S}$ that determines the active region for the subtask t,

- $\beta_t \subset \mathcal{S}$ denotes the states in which the subtask terminates,

- $\mathcal{R}_t$ is a pseudo-reward function that replaces the original reward function while planning within a subtask.

A subtask t can only stay within its active region. If an action in a subtask t transitions to a state outside $\mathcal{S}_t$ then t immediately terminates. The β_t in a subtask is akin to the deterministic version of β_o in the options framework. The pseudo-reward function assists in planning within a subtask. A basic pseudo-reward function may reward the achievement of a goal relevant to the subtask, although it has other uses also (discussed later).

Returning to our example, achieving the high-level goal of navigation may decompose into subtasks like arriving at the car location, entering the car, arriving at the destination, etc. The subtask of entering the car may further decompose into the subtasks of retrieving the car keys, opening the door, sitting in the car, and locking the door.

Notice that a task hierarchy expects far less domain knowledge than options, since it requires only a pseudo-reward function for a subtask, and not a complete policy. Thus, it is easier for a domain-designer to specify a task hierarchy compared to options. At the same time, the planner has to perform a significant amount of work, since it needs to compute an action for each active state in each subtask. We can now summarize the operation of a planner based on task hierarchies:

Input: The inputs to such a planner is an MDP and a hierarchy of subtask descriptions. To emphasize this once again, a task hierarchy planner does *not* expect policies for the subtasks to be provided.

Output: We wish to compute the policy for a subtask over the lower-level subtasks. In other words, the invocation of subtasks are the only available actions at higher levels of the hierarchy. Note that subtask selection cannot be based solely on the world state – it also depends upon which subtask the agent is in. For the same world state, the agent's policy may differ depending upon the goal of the current subtask. Hence, the planning state needs to be augmented with the current subtask id.

We define our planning state to be $[s, i]$, i.e., a pair of a world state (s) and an id for subtask t_i. The actions for each pair will be the lower-level subtasks that can be initiated.

Planning: The original algorithm to find a policy of the above form makes use of a MAXQ value decomposition [79]. Instead of computing the standard Q and V values, it computes value of a state *within a subtask*.

Definition 6.9   Value Function within a Subtask. The *value of a state s within a subtask* t_i, denoted by $V_s^*([s, i])$, is defined as the maximum expected discounted reward that can be gathered by an agent starting out in planning state $[s, i]$ (i.e., the agent is within the task t_i starting in s), such that the agent stops at the termination of t_i. Similarly, the *Q-value within a subtask*, $Q_s^*([s, i], t_j)$, denotes the maximum value within t_i if the agent chooses its first subtask as t_j in state s.

It is easy to see that values within a subtask can be expressed using the following set of Bellman equations:

$$
\begin{aligned}
V_s^*([s, i]) &= 0 & \text{(if } s \notin \mathcal{S}_{t_i}) \\
&= 0 & \text{(if } s \in \beta_{t_i}) \\
&= \max_{t_j} Q_s^*([s, i], t_j) & \text{(otherwise)} \\
Q_s^*([s, i], t_j) &= V_s^*([s, j]) + \sum_{s', N} T(s, t_j, s', N) \gamma^N \left[\mathcal{R}_{t_i}(s') + V_s^*([s', i]) \right] & (6.18)
\end{aligned}
$$

Here, $T(s, t_j, s', N)$ denotes the probability that subtask t_j started in state s terminates at s' in N steps. The N-step subtasks makes this into a Semi-MDP – hence the γ^N term (note the analog to Equation 6.14 for handling multi-step actions).

In other words, the value within a higher-level policy depends upon the reward obtained in the lower-level policy and the remaining reward (also called *completion reward*) of the same-level policy after having taken a lower-level action. It is noteworthy that the value of a lower-level subtask does not depend on the action executions of the higher-level tasks. Thus, we can run VI in a bottom-up manner, propagate the values to obtain the best actions for all active states at each level. The only issue will be to compute T. Equations analogous to 6.15 may be used to optimize this computation.

How efficient is this algorithm? If the active region of each subtask is limited then it may be quite efficient, since only a few state values may be needed per subtask. However, in practice, MAXQ decomposition becomes much more scalable in conjunction with a state abstraction for each subtask. We discuss this in Section 6.5.5.

Solution Quality: In this algorithm, the flow of information is from lower-level subtasks to higher-level subtasks only, and not vice versa. Due to this unidirectional flow of information, the decisions taken at a lower level will be independent of the higher-level goal to be achieved. It is easy to see that due to this additional restriction, the policies produced by this approach may not be hierarchically optimal. As an example, consider a navigation problem where the goal is to reach the coordinates (x, y). Let us say we decompose this into first achieving x and then achieving y. If there is an obstacle in the first subtask, the navigation will prefer the shortest way around the obstacle which may take it further away from the goal y coordinate. This will not be globally optimal but it will appear the best locally, since the subtask has no information about the global goal.

To analyze this algorithm, a weaker notion of optimality has been defined, known as *recursive optimality* [79]. The idea of recursive optimality is that the policy at each level is optimal assuming that policies of all lower levels are fixed. In general, a recursively optimal policy may not be hierarchically optimal and vice versa. A hierarchically optimal policy will always be at least as good as a recursively optimal one.

Computing a recursively optimal solution is in general more efficient than computing a hierarchically optimal solution. The comparison is akin to finding solutions in a general MDP *vs.* in an acyclic MDP. For acyclic MDPs, one can easily percolate information in a single direction without a more complicated iterative procedure.

The MAXQ decomposition has well-defined avenues for improving the solution quality. The first is the presence of the pseudo-reward function. If the programmer is able to provide the planner an accurate pseudo-reward function then a recursively optimal solution is indeed hierarchically optimal. Secondly, iterative improvement ideas [72] that re-estimate the pseudo-reward function after one complete planning phase may be adapted. These re-estimations are guaranteed to approach hierarchically optimal solutions in the limit. This approach may be especially useful when an agent needs to solve a task repeatedly – such computations can lead the agent toward an optimal solution, in due course of time.

The notions of interrupting a hierarchical solution in the middle (akin to interrupting options) may also be used to improve the quality of the solution. In the MAXQ literature, this is termed as a *hierarchically greedy execution*. In some cases, this may substantially improve the quality of the final solution.

6.5.3   HIERARCHY OF ABSTRACT MACHINES

A third representation for specifying a hierarchical structure is the *hierarchy of abstract machines* (HAM) [192].

Definition 6.10   Hierarchy of Abstract Machines. A *hierarchical abstract machine* (HAM) is defined as a finite-state automaton with each automaton node marked with one of the following:

- *Action:* Execute a primitive action in the environment.

- *Call:* Call another abstract machine as a subroutine.

- *Choice:* Non-deterministically select the next machine node.

- *Stop/Return:* Halt the execution of the machine and return to the previous call node, if any.

An abstract machine is nothing but a specific representation of a partial policy, one in which either we know what to execute, or we are at a *choice* node, in which case, we ought to decide which one of the available possibilities to choose. A *hierarchical* abstract machine has the added power to call other lower-level machines as subroutines — this leads to modular understanding and execution.

Another way to understand a HAM is that it is a partial algorithm/program that encodes our knowledge regarding various subproblems. For example, a simple way to traverse east could be go east, until there is an obstacle. If there is an obstacle, back off and go north or south a few steps, and when the obstacle is cleared go back to going east. This is easily encoded as a HAM, by having machines for going east, north, and south. The high-level machine will be for going east, and when an obstacle is seen, a *choice* node, which selects between north and south machines, will come into play.

While options create more action choices by creating more complex actions, HAM can also constrain the set of allowed actions. Where task hierarchy can only specify a high-level goal (pseudo-reward), HAM can also specify specific ground actions. Thus, HAM representation encompasses the benefits of both the previous approaches and allows for a flexible way for the domain designer to specify the domain-control knowledge. HAM is the most expressive of the three representations. With this in mind, we can describe the principles of a HAM-based planner as follows:

Input: This planner takes an MDP and a description of a HAM for it. In particular, each *action* node of each machine spcifies which action should be executed, each *call* node specifies which machine should be invoked, and each *choice* node gives the set of machines from which the planner should choose the next one for execution.

Output: The objective of the planner amounts to deciding for each *choice* node which machine to execute at that node.

Planning: Contrary to the previous section, in which we computed a recursively optimal solution, we will discuss computation of an optimal policy that respects the HAM constraints. In other words, we wish to decide hierarchically optimally which actions to choose in the choice nodes of each machine. Similar to the task-hierarchy case, this decision is not based solely on the world state, since depending upon the machine that is being run, the agent may encounter a different *choice* node. Moreover, the planner may also want to take into account the calling stack of the current machine for its decision making. Hence, the original MDP state needs to be augmented by these terms for accurate planning.

Assuming we replicate each machine multiple times, so that the HAM call graph is a tree (thus we do not need to store the call stack), the new state space for the MDP is simply $S \times H$, where H is the union of all the machine nodes. The probability function on this new state space can be appropriately defined so that, if the machine node is an action node, then both components of the state space change independently; if it is a *choice* node, then additional actions that change only the machine component are introduced, etc. It is easy to show that this new model is indeed an MDP.

Note that since a policy needs come up with decisions only for the *choice* nodes (the decisions for other node types are part of the planner input), we can reduce the MDP to transition directly to the next *choice* node. We define an operator reduce($S o H$) that achieves exactly this. As usual, the resulting decision problem is no longer an MDP — it is a Semi-MDP, as multiple time steps may elapse until the next *choice* node.

To solve the HAM planning problem, we observe that $Q^*([s, m], a)$ (m is the machine state and s is the world state) can be decomposed into multiple parts – (1) the reward obtained while the action (macro-action) a is executing (V_s), (2) the reward obtained within this machine after the completion of a (i.e., completion reward, V_c), and (3) the reward obtained after the agent exits this machine (exit reward, V_e). Thus, $V = V_s + V_c + V_e$.

Compare this equation to Equation 6.18. The MAXQ decomposition only included the first two terms. On the other hand, HAM planners include the third term, the exit rewards. Solving this equation returns a hierarchically optimal solution, since the exit term is encompassing information flow from the top levels to the bottom levels.

Building a HAM planner involves writing dynamic programming equations for each of the three terms. Value Iteration-based algorithms simultaneously compute all components to obtain the optimal Q-values and find a hierarchically optimal policy. These equations adapt Equation 6.18 to HAMs and extend them with the V_e term. Please refer to the original paper [5] for more details. VI-like algorithms can be used for computation of the hierarchically optimal value function and thus a hierarchically optimal policy.

Solution Quality: As presented, HAM algorithms compute a hierarchically optimal policy. An astute reader may observe that one may speed up HAM planning by employing ideas from MAXQ value decomposition and recursive optimality at the loss of hierarchical optimality. Alternatively, one may use the ideas from this section to add the exit term to MAXQ decomposition and get a hierarchically optimal solution for the task hierarchy case.

6.5.4 OTHER APPROACHES

Modern hierarchical planners (using options, hierarchies of abstract machines, or task-hierarchies) are built on top of earlier research in hierarchical learning. The work on hierarchical learners dates back to at least 1952, when Ashby developed a gating mechanism to handle recurrent situations and repetitive environmental disturbances [6]. His mechanism mapped environmental stimuli to triggers aimed at preserving equilibrium of the system. A few years later, he also came up with the idea that this gating mechanism need not stop within two levels [7]. Indeed, early work in hierarchical planning aimed at building a gating mechanism that switches between known behaviors [161].

A subsequent system proposed a feudal representation [67], in which a hierarchy of managers and submanagers controlled an agent. The aim of the feudal planner is not to *pick* a submanager, but instead to give an appropriate *command* to that submanager. In fact, each level has exactly one manager. All the managers learn the values of possible commands they can give to their submanager, given the current state and the command given to them by their own manager. At the lowest level, a command is just a primitive action.

Hierarchical DYNA learns elementary behaviors that achieve recognizable subgoal conditions, which are then combined at a higher level by learning a suitable gating function to achieve the overall goal [218]. For example, suppose the overall goal could be to traverse to A and B. Elementary subtasks could be to traverse to these locations individually, and a gating mechanism might need to learn the sequence "visit A and then visit B" to achieve the goal. Thus, contrary to feudal learning, a manager has multiple submanagers and it has to switch among them.

The *Hierarchical Distance to the Goal* (HDG) algorithm divides the available state space into regions [122]. Each region is associated with a landmark. The distance to the goal is decomposed in three components — distance of the start state to the nearest landmark, distance between that landmark to the landmark closest to the goal, and distance between that landmark to the goal. Learning is applied in each of these steps for all possible initial states and goal states. The HDG algorithm was later extended to multi-level hierarchies [185].

Another key idea developed in this work is the notion of non-hierarchical execution. We can observe that the agent need not visit each landmark in order to reach the goal — it could just direct itself toward the current landmark. If it is beneficial to start moving to the next landmark without visiting the current one, then it would do so. In other words, the policy executed would not be strictly hierarchical. The hierarchy would be used only in learning and decomposing the value function, and the current greedy action based on that decomposition would be executed. This idea is also present in modern systems, e.g., when terminating an option early or during a hierarchically greedy execution in the MAXQ framework.

An alternative approach to problem decomposition is to subdivide an MDP into multiple sub-MDPs for different regions of the state space [72]. The policies for each sub-MDP are combined to solve the higher-level problem. This work also introduced an iterative mechanism that recomputed the sub-MDP policies and higher-level policy, and iterated through them, until no more improvement was observed.

In relatively more recent work, researchers have extended probabilistic planning to use other ways to specify domain-control knowledge — hierarchical task networks (HTNs) and control formulas [8; 188]. These approaches extend MDP algorithms for use with such domain knowledge [141; 227].

The AO* algorithm has also been extended to hierarchical planning [180]. Hierarchical AO* exploits the hierarchy to delay the backups in AO* and obtains computational savings over AO*. To our knowledge this is the only work that combines hierarchies with heuristic search for MDPs.

Finally, a different kind of problem decomposition known as parallel decomposition has also been studied. In this paradigm, the overall MDP is a cross-product of the component sub-MDPs [98]. Such decompositions are most useful for modeling multi-agent problems. We briefly discuss them in Section 7.2.2.

6.5.5   STATE ABSTRACTION IN HIERARCHICAL MDP

As discussed earlier, state abstractions and hierarchical decomposition of the problems often go hand in hand. A state abstraction is a specific aggregation technique based on ignoring state variables. An abstract state space selects a set of fewer state variables than in the original problem. An abstract state is an assignment to that subset of state variables. Therefore, many world states are mapped to one abstract state.

The idea of a hierarchical structure over state space was made popular by Knoblock [127] when he proposed an automated mechanism to divide the state variables hierarchically. The basis of this division was domain analysis for finding higher-level variables that remained unaffected while modifying the lower-level ones. Thus, having discovered the hierarchy, planning could be conducted hierarchically top down, level by level, using a backtracking search.

A hierarchical state abstraction accompanying a hierarchical problem decomposition is quite useful. Recall that most MDP based algorithms are exponential in the number of state variables. Thus, dropping even a few state variables from a (sub)problem leads to significant reduction in the planning times.

Several kinds of state abstractions are used in hierarchical planning [79]. Probably the most popular abstractions are based on *relevance*. A subtask may have only a few relevant variables. By abstracting to only those variables, problem sizes may be drastically reduced. This form of abstraction applies to the current task and all its subtasks.

Another form of abstraction is based on *funneling*. A funnel subtask is one that maps a larger number of states into a small number of states. Thus, the completion reward of a funnel action can be represented using fewer state variables. Funneling helps in abstracting states for the higher-level subtasks.

Yet another form of abstraction is called *shielding*. If the termination condition of an ancestor task causes the subtask to be reachable only in a subset of states, then the other states can be abstracted away.

These three and others are common conditions used by domain designers to specify state abstractions. In conjunction with such abstractions hierarchical planners often run much faster than the flat counterparts.

There is some work on automatically identifying state abstractions. For a non-hierarchical MDP, the main principle is to automatically cluster states so that the rewards and transitions between clusters for each state in a cluster are the same. In other words, all similarly behaving states get clustered into a single macro-state [69]. The same principle may be utilized for a hierarchical MDP. Each subtask often has a much smaller number of active actions. Thus, even if the flat MDP does not have many state clusters, different subtasks commonly do [194].

Other research has used topological analysis of the state space to cluster states into connected components. These clusters form the higher-level states [11]. Some approximate approaches have also been developed; for example, a set of relevant state variables for each task is determined through statistical tests on the Q-values of different states with differing values of the variables [164].

6.5.6   LEARNING HIERARCHICAL KNOWLEDGE

All the hierarchical MDPs discussed so far have assumed pre-specified hierarchical knowledge about a domain. However, this puts an additional burden on the domain designer. Moreover, it does not satisfy the vision of an autonomous AI agent, in which the agent is intelligent enough to perform planning from just the basic domain description.

Although discovering high-quality hierarchies automatically is still very much an open question, there exists a body of work in this area. We are not aware of any work that learns HAMs, but for learning options a common approach is to identify important subgoals in the domain and then compute an option policy for each subgoal. On the other hand, for a task-hierarchy, we need to obtain a hierarchical subtask decomposition, where the pseudo-reward is often based on the subgoal for each subtask.

In other words, both options and task hierarchies are learned by first identifying subgoals. The key question, then, is how to identify important subgoals in a domain. Several methods have been proposed in the literature. An early method tried to achieve a conjunctive goal, one literal at a time. Of course, this may not always simplify the planning problem (for example, when two literals can only be achieved together).

Probably the largest number of approaches for subgoal discovery use the notion of a *central* state. These central states have been defined in various ways, e.g., as states that is typically found in successful trajectories but not in unsuccessful ones [174], states that border strongly connected areas [178], and states that allow transitions to different parts of the environment [215]. These ideas have been generalized by using a graph-centrality measure of *betweenness* to identify such subgoals [216].

Other algorithms define the subgoals around the changing values of state variables. HEXQ [109] relies on the differences in the frequencies of value changes in state variables to determine the task-subtask relationships. The intuition is that the frequently changing variable will likely be attached to the lowest-level subtasks and vice versa. The VISA algorithm [118] uses similar ideas but, instead of the frequency of change, performs domain analysis to infer causal relationships between variables. It clusters the variables so that there is acyclic influence between the variables in different clusters. This naturally defines a task-subtask hierarchy, where the state variables which influence the values of other state variables are assigned to lower levels.

Another set of approaches is based on analysis of successful trajectories. Search over the space of hierarchies has been used to fit the successful trajectories [164]. HI-MAT [177] builds on this by causally annotating the trajectories and recursively partitioning them to find the subgoal hierarchy. Notice that this requires successful trajectories, which may only be achieved when the first problem is solvable by non-hierarchical techniques. Therefore, such approaches are most common in transfer learning literature, where hierarchy from one domain is transferred to another.

6.5.7   DISCUSSION

The need for hierarchical planning can be motivated in a variety of ways. It can be seen as an effective method for managing the curse of dimensionality, especially in conjunction with state abstraction. Solving multiple problems of smaller sizes is much faster than solving one larger problem. A hierarchical representation is an intuitive way for a domain designer to provide additional domain knowledge, which can then be exploited for faster planning. Hierarchical subtasks often encode important behaviors or skills in a domain. These behaviors become important building blocks for higher level planning and can be used in a variety of problems. For example, the navigation skill of a robot may be used to create a map of the building and also to fetch printed sheets from the printer room. For these reasons, hierarchical problem solving is an active area of research within probabilistic planning.

An important question is how to choose an appropriate representation. Options, HAM, and task hierarchies can be understood to lie on a continuum in terms of the amount of domain knowledge expected from a designer. The two extremes are options, which are complete pre-specified policies and task hierarchies, which require a domain designer to provide just a subgoal decomposition. HAMs is the most expressive representation and allows a designer to provide as much knowledge as they may have. Thus, the tradeoff is in how much knowledge the designer provides and how much work is done by the planning algorithm.

Options are simple to implement and are effective in defining high-level skills reusable in different problems. Common option algorithms (those that use a union of options and the primitive actions for planning) do not simplify the planning task, because they augment an MDP. Moreover, options do not have an explicit notion of task decomposition. HAMs can be difficult to design. On the other hand, they do restrict the space of policies and make planning tasks easier. Task hierarchies are easier to design and weakly restrict the space of solutions.

Both HAMs and task hierarchies have similar algorithms, and the choice there is in which notion of optimality and, hence, which value decomposition to use. The hierarchically optimal version of the algorithm may provide better solutions, but may not be as fast. The recursively optimal version may provide lower-quality solutions efficiently and would also allow easy reuse of the same sub-policy for solving the larger task.

Learning hierarchical structures and associated state abstractions automatically is a holy grail of AI. This would be a clear demonstration of intelligence, since the agent would uncover the hidden structure in a domain to achieve a modular understanding and efficient decision making. We have discussed some research in this area, but it contains many challenging problems for which no widely successful general solutions exist yet. In our opinion, the problem of hierarchy discovery deserves more research attention; a breakthrough in it would likely have a large impact on domain-independent MDP algorithms.

6.6   HYBRIDIZED PLANNING

In addition to the standard online and offline scenarios, researchers have studied a third setting known as *anytime planning*. Anytime planners run in offline mode, and are expected to achieve better and better solutions as more computation time is made available. Usually anytime planners find a first solution fast and use the remaining computation time to successively improve the solution. In this section, we describe an anytime planner that is a hybrid of an optimal MDP solver and a fast suboptimal planner.

Hybridizing two planners to achieve an anytime planner is a general idea and may be used with different combinations of planners. For example, we may use LAO*, LRTDP or their extensions for the optimal planner. A natural choice for a fast suboptimal planner may be a determinization-based algorithm, e.g., FF on an all-outcome determinization. Because the determinized problems are far easier to solve than the original MDPs, a first solution may be found very quickly. On the other hand, their key drawback is that the uncertainty in the domain is compiled away, which may result in catastrophic action choices, rendering the goal unreachable in the face of undesired outcomes.

We present HybPlan, whose suboptimal planner also relaxes the original MDP, but not to a deterministic domain; instead, it uses a non-deterministic relaxation of the MDP. Before we describe the algorithm we point out that HybPlan was originally developed in the context of SSP_{s_0} MDPs, i.e., MDPs with a proper policy closed with respect to s_0. That said, the idea behind hybridization is general and may be extended to other MDPs also.

Non-deterministic planning is a formalism that allows for uncertainty in action's effects, but represents it as a disjunction over possible next states, without associating any cost or probability information with the actions. Any MDP can be converted into a non-deterministic problem by stripping away costs and probabilities, and allowing all non-zero probability successors in an action's disjunctive effect. For example, if an action a in state s leads to states s_1 and s_2 with probabilities p and $1 - p$, then the non-deterministic domain will represent the successor of (s, a) as the set $\{s_1, s_2\}$.

We define a *strong-cyclic* solution to a non-deterministic planning problem as an execution hypergraph rooted at the start state s_0, which reaches the goal while admitting loops, but disallowing absorbing goal-less cycles (cycles in which goal-achievement is impossible) [54]. A strong-cyclic solution is guaranteed to reach the goal in a finite number of steps, even though one may not be able to bound the number. Computing a strong-cyclic solution is easier than computing the optimal MDP policy, since the former is looking for just one solution, instead of optimizing from the space of solutions. Also notice that, even though potentially suboptimal, a strong-cyclic solution is a *proper* policy for the MDP.

A simple algorithm will be to directly use the solution from the non-deterministic problem. This will take the agent toward the goal, even though the actions may be costly. In case additional planning time is available, we can try to improve this solution.

HybPlan is a hybridized planner [167] that combines the policy returned by a non-deterministic planner (e.g., MBP[3] [54]) with partial computations of LRTDP to obtain a strong anytime behavior — stopping HybPlan before LRTDP's convergence still results in high-quality proper policies.

The key idea of the algorithm is quite simple. It runs LRTDP for some number of trials and then accesses the current greedy partial policy from LRTDP. This policy may be quite suboptimal since LRTDP has not converged yet, and might not even be defined for many relevant states (those that have not yet been visited by LRTDP). HybPlan tries to complete this partial policy by adding guidance from MBP's solution for states that are not explored enough by LRTDP.

Even this hybridized policy may not be proper, since LRTDP may be suggesting bad actions due to its incomplete computation. HybPlan then tries to identify states with bad actions and backs off to MBP's actions for those states. This process continues until the whole policy is deemed proper.

HybPlan's strength is its anytime performance. By performing regular intermediate computations of hybridized policies and storing the best policy found so far, the algorithm may find a near-optimal solution quite soon; although, HybPlan would not be able to prove optimality until LRTDP's convergence. HybPlan makes good use of the available time — given infinite time, HybPlan reduces to LRTDP (plus some overhead) and given extremely limited time, it behaves as well as MBP. However, for the intermediate cases, it exploits the power of hybridization to achieve policies better than both.

HybPlan can also be run in a non-anytime mode in which it terminates with a desired quality bound. This is achieved by finding the gap between LRTDP's current value $V(s_0)$ (which is a lower bound to the optimal value, if the initializing heuristic is admissible) and the value of intermediate hybridized policy (which is an upper bound). The difference in the two values gives us an absolute error of the current policy. HybPlan can stop when this error is less than the desired bound.

HybPlan may be understood in two ways. The first view is MBP-centric. If we run HybPlan without any LRTDP computation then only MBP's policy will be returned. This solution will be

---

[3]MBP is a BDD-based solver that tries to construct a strong-cyclic solution by backward chaining from the goal and removing bad loops, i.e., those for which no chance of goal achievement exists. For more details please see the original paper [54].

proper, but quite possibly low quality. HybPlan successively improves the quality of this basic solution from MBP by using additional information from LRTDP. An alternative view is LRTDP-centric. We draw from the intuition that, in LRTDP, the partial greedy policy improves gradually and eventually gets defined for all relevant states accurately. But before convergence, the current greedy policy may not even be defined for many states and may be inaccurate for others, which have not been explored enough. HybPlan uses this partial policy and completes it by adding in solutions from MBP, thus making the final policy proper. In essence, both views are useful: each algorithm patches the other's weakness.

Finally, we add that HybPlan is just one instance of the general idea of hybridizing planners – taking solutions from a fast but suboptimal algorithm and fusing it with a partial solution from a slow but optimal algorithm [165]. A few other hybridizations have been suggested in the original paper [167], though none have been implemented. It will be interesting to hybridize the determinization-based algorithms with pure MDP solvers, to obtain a possibly better use of available time.

6.7   A COMPARISON OF DIFFERENT ALGORITHMS

The large variety of available approximation algorithms brings up a natural question: which one is best in a given set of circumstances? This question is nearly impossible to answer in general, as each of the algorithms we have mentioned has been compared to most others in very few settings, if at all. However, the comparisons that have been made do allow us to provide some guidance regarding the choice of a planner for a specific task.

Perhaps no other criterion determines the appropriateness of a given planning algorithm more than the amount of planning time available before and during the policy execution. If the agent does not have the opportunity to plan in advance and has to act according to circumstances, the choice is restricted to online planning algorithms. These techniques typically have a way of coming up with a "quick and dirty" solution, possibly at the cost of overlooking some negative consequences of the chosen decision. The epitome of online planning is FF-Replan. By far the fastest of the algorithms we discussed, it fails to consider what happens if its chosen plan of action goes awry. FF-Hindsight largely fixes FF-Replan's deficiencies, but is comparatively much slower as a result, since it has to call the deterministic planner many more times.

If the agent can afford to "think in advance," i.e., plan offline, the range of planner choices for it is much broader. If additional domain knowledge can be specified, either in the form of basis functions or a hierarchical problem decomposition, then it can be leveraged via API or hierarchical planners, as appropriate. These algorithms can perform very well, scaling to larger problems, producing near-optimal and, importantly, human-interpretable policies. The drawback of these approaches is that they put additional burden on the human expert to hand-craft the additional input. If circumstances allow for offline planning but can cut it short at any moment, HybPlan is one of the most appropriate tools due to its good anytime behavior.

If the algorithm needs to be installed on a completely autonomous system, ReTrASE is a viable option. Its main drawbacks are the lack of a stopping condition or any guarantees on the output policy quality, both of which can be a significant impediment for autonomous system operation. FPG's strengths and weaknesses are similar to ReTrASE's. One of its advantages is that it can operate in partially observable environments without much modification [49]. On the downside, its convergence may be very slow.

Of course, there are many scenarios where the agent can combine the best of both worlds — it has the time to plan both before and during policy execution. RFF and UCT are very well suited for these settings. RFF, being essentially an incremental contingency planning algorithm, lets the agent build a skeleton of the policy offline. The agent can then easily augment the skeleton as necessary while acting. UCT similarly allows the agent to get a feeling for reasonable behaviors in different states before commencing with policy execution, and to improve its knowledge of the state space whenever it has the time to plan more. The choice between RFF and UCT is one of convenience of use. Both algorithms have several parameters to adjust. However, RFF's parameters are fairly easy to tune. In addition, their values generalize to many problems. UCT, on the other hand, needs a lot of experimentation with parameter values (e.g., the weights of the initial heuristic values assigned to various state-action pairs) to show its best. Moreover, its parameter settings often fail to generalize even across different problems within the same domain. At the same time, UCT is the more natural alternative for finite-horizon and infinite-horizon discounted-reward MDPs.

Moreover, UCT's model-free nature gives it an edge in scenarios where the model is not fully known or cannot be used explicitly. This is the case in scenarios where at least one action may cause transitions to a very large number of states. Computing the Q-values for this action in closed form is very expensive, but UCT circumvents this difficulty via its use of Monte Carlo sampling.

CHAPTER 7

Advanced Notes

So far, to keep the discussion focused, our algorithms have made several assumptions about the MDP representation, such as discrete state and action spaces, flat action space, etc.; assumptions about problem structure, such as all improper policies assumed to be infinite cost; and assumptions about our knowledge about the domain, such as access to full MDP structure and parameters in advance.

Indeed, much research has gone beyond these simplifying assumptions and has investigated many complex extensions to the MDP model. The strength of MDPs is its robust theory, which is applicable to many of these extensions. Unfortunately, naive extensions of the existing algorithms usually suffer computational blowups. As a result, much of the research focuses on managing and reducing the computation.

Space constraints for the book, along with the large breadth and depth of this research makes it impractical for us to do justice to this literature. In this chapter, we take a sweeping look at a few important advances. Our goal is not to be exhaustive but to give a flavor of the key ideas and the pointers from which a reader may launch a more focused advanced study.

7.1 MDPS WITH CONTINUOUS OR HYBRID STATES

We first relax the assumption of discrete state spaces and allow continuous variables in the domains, although still assume actions to be discrete. This leads to two MDP formulations, Continuous MDPs (when all variables are continuous) and Hybrid MDPs (which have a mix of discrete and continuous variables).[1] This formulation has several important applications. Domains such as Mars rover planning require us to keep track of location and battery power, which are continuous variables [43]. Any planning of daily activities is based on the available time, which is a continuous variable [41].

The Bellman equations are easily extensible to these situations – replace the sum over next states by an integration over next states; replace the probability of next state by the probability density function of the next state. This leads to a form of Bellman equation (analogous to Equation 3.3) for a Continuous MDP ($\mathcal{T}^{pdf}$ refers to the transition function represented as a probability density function):

$$V^*(s) = \min_{a \in \mathcal{A}} \int_{s' \in \mathcal{S}} \mathcal{T}^{pdf}(s, a, s') \left[\mathcal{C}(s, a, s') + V^*(s') \right] ds' \qquad (7.1)$$

---

[1] Hybrid MDPs are also studied as Discrete-Continuous MDPs (DC-MDPs) in literature [209].

For a Hybrid MDP, the state space $\mathcal{S}$ can be assumed to comprise two sets of state variables: $\mathcal{X}$, the discrete state variables $\{X_1, \ldots, X_k\}$ and $\mathcal{Y}$, a set of continuous variables $\{Y_1, \ldots, Y_k\}$. We use (x, y) to denote a state s, which has x and y as discrete and continuous components respectively. Similarly, we may factor the transition function into discrete and continuous components. This formulation results in the following Bellman equation variant [166]:

$$V^*(x, y) = \min_{a \in \mathcal{A}} \int_{y'} \mathcal{T}_{x'}^{pdf}(x, y, a, y') \left(\sum_{x'} \mathcal{T}(x, y, a, x') \left[\mathcal{C}(x, y, a, x') + V^*(x', y') \right] \right) dy'$$

(7.2)

Notice that the transition function for these MDPs is continuous (or hybrid), and amidst other things, needs to also represent the probability density function of the new values of the continuous variables. In the equation above, $\mathcal{T}_{x'}^{pdf}(x, y, a, y')$ represents the probability density of new state's continuous component (y'), when action a was taken in state (x, y) and the new state has discrete component x'. Other functions, such as costs, are also continuous distributions.

It is interesting to note that while a discrete MDP may be represented as a flat MDP, a continuous or a hybrid MDP necessitates a factored representation. Since the number of states are uncountably infinite, they cannot be possibly represented in a flat form.

Representation-wise the transition function of a continuous variable is often defined relative to its previous value. That is, $\mathcal{T}^{pdf}$ maintains the probability that the variable will *change* by a given amount. For instance, if the rover is navigating to another location, pdf will represent a density value for the amount of energy spent in the navigation action. The transition function is computed by subtracting the actual energy spent from the current battery available.

We can define value iteration and heuristic search extensions based on these Bellman equations, except that all value functions are now continuous (or mixed) functions and, therefore, cannot be represented as tables directly. The first challenge in solving these problems is in defining compact representations for these continuous functions. A related problem is efficient computation of value updates.

7.1.1 VALUE FUNCTION REPRESENTATIONS

We start by discussing Continuous MDPs. An obvious first idea is to discretize the continuous part of the state space as well as the transition function [53]. This converts the whole problem into a discrete problem and previously covered techniques apply. The level of discretization may be decided by the available memory and the desired accuracy of the solution. The transition function is computed by replacing each (continuous) successor with the nearest-neighbor discretized state, or by encoding a stochastic transition to several nearest neighbors, with probability dependent on the distance from these neighbors. Similar ideas are employed when executing a discretized policy from an intermediate state not represented in the discretized MDP.

A natural drawback of this approach is that it blows up the state space (and transition function) enormously, making the algorithms highly inefficient. That said, the simple discretization approach is still popular as a first step and is a useful technique to solve problems with a small number of state variables [64; 65].

Researchers have also proposed adaptive discretization ideas. Under these approaches a coarser discretization is successively refined to obtain a solution at the desired accuracy. One technique successively solves an MDP at a certain discretization level and then refines the discretization locally [186].

An alternative is to solve the MDP only once but use more structured representations to encode the value function. A data structure that implements this is called *Rectangular Piecewise Constant* value function representation (RPWC). It partitions the continuous state space into hyper-rectangles (each rectangle's boundaries defined by lower and upper bounds on the values of each variable). Exactly one value of the function is specified for one hyper-rectangle. A *kd*-tree [55] can be used to store an RPWC function.

How to compute a Bellman update with values represented as RPWC? The simplest solution is to approximate by discretizing the action transitions. We can prove that under this setting, if V_n was represented as an RPWC, then after a Bellman backup, V_{n+1} is also representable as an RPWC [84]. However, V_{n+1} may require more hyper-rectangles than V_n. As a general principle, we can implement a VI-based algorithm over a representation easily if we can guarantee that V_{n+1}, computed via Bellman backup over V_n, uses the same representation as V_n. Notice that this is analogous to Chapter 5 where ADDs satisfied this property and lent themselves to easy implementations of VI, LAO*, etc. In essence, algorithms using RPWC may be thought of as automatically finding an appropriate level of adaptive discretization, since states with similar values get clustered together.

Instead of discretizing, we could use RPWC to represent the pdfs of the continuous transitions also. In that case, V_{n+1} is no longer in RPWC, but we can further approximate it and represent it back in RPWC. This is known as *lazy discretization* [150].

One of the drawbacks of RPWC is that it assigns a constant value for each hyper-rectangle. So, if the state values had non-constant dependencies on continuous variables, (e.g., value was linear in time elapsed), it would require an infinite number of pieces to represent them. Or, in other words, the function will be discretized to its lowest allowed discretization level, giving us no benefit of the structured representation. Since linear costs are common, researchers have studied RPWL – rectangular piecewise linear value function representation [84; 159; 187; 193; 196]. The simplest version of this is when there is only one continuous variable, say, time [41].

The computational ideas for RPWC also extend to RPWL. For example, if V_n is in RPWL, and if action transitions are discretized then (under additional restrictions), V_{n+1} is also in RPWL. For lazy discretization secondary approximations are needed as before.

Those familiar with Partially Observable Markov Decision Processes (POMDPs, see Section 7.5) literature may find connections between Continuous MDPs and POMDPs. Indeed, a POMDP is just a Continuous MDP with a state variable for each world state. The ideas of Piecewise Linear Value Functions were originally developed in POMDP literature, and later adapted to Continuous and Hybrid MDPs [123].

Other domains may be most naturally solved using representations that are not piecewise linear or constant. For instance, there have been attempts to represent transition functions with a mixture of Beta distributions [107], or exponential distributions [144; 162]. State value representations using piecewise one-switch functions [158; 160], piecewise gamma representations [162], and a combination of basis functions [145; 146] have been explored. In special cases, some of these combinations, e.g., phase-type distributions with piecewise-gamma functions for values of a state, can lead to analytical solutions for computing Bellman backups.

We now direct our attention to Hybrid MDPs. Recall that the state space in a Hybrid MDP has two components – continuous and discrete. To solve these, with each discrete (abstract) state[2] we associate a complete value function (for all possible values of continuous variables) [166]. Thus, a full RPWC or RPWL function is associated with each possible discrete component of the state. All other ideas extend analogously.

Researchers have also begun exploring extensions of ADD representations to continuous domains, called Extended ADDs (XADD). In these XADDs, each internal node, which represents a way to split the value function, can be any inequality over continuous variables or discrete variables [209]. RFF algorithm has also been extended to hybrid MDPs [230]. Hybrid RFF (HRFF) uses hierarchical hash tables to represent the hybrid value functions.

Finally, LP-based approaches also naturally extend to Continuous and Hybrid MDPs [107; 145]. The algorithm, known as Hybrid Approximate Linear Programming (HALP), extends the ALP ideas and can handle several function representations, e.g., piecewise linear, polynomial, and beta functions. To make the algorithms efficient, the LP is relaxed by maintaining only a subset of the original constraints. Several ways to relax the LPs have been studied, e.g., sampling of constraints, grid discretization of constraint space and adaptive search for violated constraints [99; 107; 142; 143].

7.1.2   HEURISTIC SEARCH FOR HYBRID MDPS

Hybrid AO* [166; 179] is a heuristic search algorithm for MDPs with hybrid state spaces and a known initial state. As before, it attempts to explore only the part of the state space that is relevant to finding a policy from the start state. However, there is one key difference. We wish to extend the reachability ideas to the continuous component of the state space too. In other words, for each discrete component, parts of continuous component may be unreachable or provably suboptimal, and this information needs to be incorporated into the algorithm.

Hybrid AO* performs search in the discrete part of the state space as before, but instead of one, it stores two (continuous) functions with each abstract (discrete) state. One is the current value function (V_n) and the other is the probability density function over the possible values of the continuous variables. For example, some values in the continuous component (like high battery availability) may not be possible in an abstract state where the rover is at a faraway location. This is because a significant amount of energy may have to be spent in reaching that location. Thus,

---

[2]Since just the discrete component is not a full state, we call it an *abstract* state.

exploring this abstract state (and computing its value) with a very high available battery power is wasteful computation.

Recall that the original AO* algorithm (Section 4.3.4) operates in two steps: a forward expansion to expand the fringe states, and a backward pass to update values of the ancestors of the expanded states. Hybrid AO* operates in three passes. The first and second passes are similar to AO*'s. Hybrid AO* adds a third pass to propagate forward the reachability information of the continuous part of the state space. Hybrid AO* with structured value functions enormously speeds up the overall computation. It was implemented in a research prototype for the next generation Mars rover for planning its daily activities [166; 179].

Search-based ideas for solving Hybrid MDPs are also discussed in other algorithms, e.g., in the Dynamic Probabilistic Function Propagation algorithm [163]. But overall, this field can make use of more researcher attention. Extensions to Hybrid LAO*, RTDP, and other creative ways to manage computation will be very useful for several real scenarios that are modeled as Hybrid MDPs.

7.1.3 CONTINUOUS ACTIONS

MDPs are also extended to include actions that are parameterized by continuous variables [145], e.g., in the Extended MDP (XMDP) formalism [199] and the Continuous State and Action MDP (CSA-MDP) model [257]. Continuous actions can represent actions such as "drive north for m meters" or "refuel for t time." Of course, the min over all actions needs to be replaced by an infimum over the continuous space of parameterized actions for the Bellman equations. An approach to solve such MDPs performs symbolic dynamic programming using XADDs [257].

The problems with continuous action spaces are most commonly studied in the control theory, where a controller is trying to control a physical system, such as a motor, joints of a robot, etc. Here, the transition of the system is often described by differential equations; in the case of physical systems, they express the laws of physics.

A very different set of ideas is used to efficiently compute Bellman backups, for instance, the use of Hamiltonian-Jacobi Bellman equations in solving such MDPs. This body of literature is beyond the scope of our book, but see [148; 222] for an introduction.

7.2   MDP WITH CONCURRENCY AND DURATIVE ACTIONS

Standard MDPs make two common assumptions – (1) only one action is executed at each decision epoch, and (2) actions take unit time and their effects are already known to the agent at the next decision epoch. These are fine assumptions for developing the model and algorithms, but they break down in several real settings. For instance, actions typically take differing amounts of time – often the durations may be probabilistic too. Moreover, in the case of multiple agents or even in the case of a single agent with multiple effectors concurrency of actions becomes important. A system of elevators, a grid of traffic lights, a robot with multiple arms are all common examples of scenarios with concurrent actions. In these situations it is important to extend MDPs to handle concurrency and durative actions. We first start by MDPs that relax only one of these assumptions, and finally discuss models relaxing both.

7.2.1   DURATIVE ACTIONS

The simplest durative extension to an MDP involves non-concurrent actions that take deterministic durations. As long as the next decision epoch is when the previous action completed, this does not complicate the SSP MDP model or algorithms. If, additionally, the goal has a deadline, then modeling needs to incorporate the current time in state space and can be dealt in a manner similar to Hybrid MDPs of the previous section.

In the case of discounted problems, however, the durative actions affect the discounting process, since the rewards obtained after a longer action need to be discounted more. This model, known as *Semi-MDPs* (recall the discussion in Section 6.5) [115; 197], results in the following Bellman equations, with $\Delta(a)$ being the duration of the action:

$$V^*(s) = \max_{a \in \mathcal{A}} \sum_{s' \in \mathcal{S}} \mathcal{T}(s, a, s') \left[\mathcal{R}(s, a, s') + \gamma^{\Delta(a)} V^*(s') \right] \tag{7.3}$$

In these equations we assumed that action durations are state-independent and deterministic. Semi-MDPs can naturally handle duration uncertainty by allowing a probability distribution over durations, and taking an additional expectation over them in the Bellman equations. They can also handle state-dependent durations. Once the Bellman equations are defined, the MDP algorithms are easily extended to Semi-MDPs [42]. Commonly, a special case of Semi-MDPs is studied, which restricts the durations to be exponential distributions. This model is known as a Continuous-Time Markov Decision Process [197].

Durative actions often allow for action parameterizations, e.g., in modeling actions like "wait for t time units." If time is treated as a continuous variable, this falls under Semi-MDPs with continuous actions. If time is also needed in the state space (e.g., for a domain with deadlines on the goal achievement), it will result in a model involving continuous states, actions, and action durations – probably too complex to be solvable for practical problems yet.

Several algorithms have been developed for problems with durative actions. A few notable examples are an adaptation of standard MDP algorithms like RTDP [42], a VI-based algorithm that outputs piecewise linear value functions [41], use of simple temporal networks for temporal reasoning [92] and incremental contingency planning, which successively adds branch points to a straight-line plan [43; 182].

7.2.2   CONCURRENT ACTIONS

Concurrent MDP (CoMDP) extend MDPs to concurrent actions, when all actions take unit time [173]. At each time step, a CoMDP executes a *combo* – a subset of actions. CoMDPs assume that the effects of all actions are available at the next time step. Since these actions can potentially conflict with each other, say, by modifying the same state variables, a CoMDP also studies the general principles under which a combo of actions is safe to apply. Researchers have extended this to problems where actions may conflict [56].

A CoMDP, after defining the joint transition function and cost function, is essentially an MDP in a potentially exponential action space. This is because there are at most $2^{|A|}$ action combos. This leads to an exponential blowup in each Bellman backup. To manage the computation, ways to prune suboptimal action combos (similar to action-elimination theorem from Section 4.5.1) have been studied. Approximate algorithms subsample the set of action combos in each backup, which often leads to near-optimal solutions [169]. MBFAR algorithm extends symbolic dynamic programming for such problems [201].

Multi-agent scenarios are the most important application of Concurrent MDPs, since each agent can execute an action in parallel with actions of other agents. Approximate linear programming (ALP) has been extended to this special case [101]. Hierarchical decomposition of the models is also natural with multiple agents – there is a separate sub-MDP for each agent [98]. In these cases, the value function is as usual defined as a linear combination of basis functions. Additional assumptions restricting basis functions to depend only on a few agents reduce the size of the LP further.

Factorial MDP is a special type of MDP, which represents a set of smaller weakly coupled MDPs. The separate MDPs are completely independent except for some common resource constraints [181; 217]. These weakly-coupled MDPs are solved independently and merged later. The possible concurrency of actions from different sub-MDPs is a by-product of this approach.

Finally, *Paragraph* [153] formulates the planning with concurrency as a regression search over the probabilistic planning graph [27]. It uses techniques like nogood learning and mutex reasoning to speed up policy construction.

7.2.3   CONCURRENT, DURATIVE ACTIONS

Planning problems become significantly harder when the domains have concurrent and temporal actions. These are now out of the purview of Semi-MDPs or Concurrent MDPs. A model that can successfully handle both of these is known as *Generalized Semi-Markov Decision Process (GSMDP)*. This is strictly more expressive than the previous models, and it is also able to incorporate asynchronous events (e.g., a machine going down by itself).

Extending VI-based algorithms to concurrent probabilistic temporal planning (CPTP) problems results in two combinatorial blowups – state space explosion and decision-epoch explosion. First, the decision about the next action not only depends on the current world state, but also on the events expected in the future (e.g., the termination of currently executing actions). This implies that one needs to augment the state space to incorporate this information when taking the decisions [170; 173].

Second, depending on the domain, it may also result in decision-epoch explosion. Recall that Semi-MDPs assumed decision epochs to be at action completions. This may no longer be sufficient (optimal) in the face of concurrency.

As an example, consider an action that may terminate in 1 min or may take 5 mins to terminate with equal probabilities. If we started it, and after 1 min realized that the action has not terminated, we know that we need to wait another 4 mins for this action. This new information may lead us to start another concurrent action. Moreover, notice that this new decision was taken at a non-event, i.e., when no action completion happened. This suggests that our decision epochs could be intermediate points in an action execution. This blows up the decision-epoch space enormously. Whether we can restrict our decision epoch space without sacrificing optimality is an open research question to our knowledge [171; 172].

By restricting the decision epoch space, augmenting the state space, and treating time as discrete, we can convert such problems into a Concurrent MDP. Informative heuristics, hybridized approximation algorithms, and determinization of uncertain durations are some techniques to speed up the solution algorithms [171; 172; 173]. Planning graph-based heuristics have also been explored in a planner called Prottle [152].

FPG [1; 50] is another popular algorithm, which learns a set of neural networks modeling a stochastic policy – for each action, a different neural network computes the probability of executing it based on state features (see Section 6.4.3). In the execution phase the decision for an action, i.e., whether the action needs to be executed or not, is taken independently of decisions regarding other actions. In this way FPG is able to effectively sidestep the blowup caused by exponential combinations of actions. In practice it is able to very quickly compute good concurrent policies.

These models have been applied to several military operations planning scenarios. In one scenario, search in an augmented state is performed with domain-dependent heuristics [2] and in another, target-specific plan space searches are carried out and multiple plans merged to achieve a global solution [19].

The GSMDP (the model that also allows asynchronous events) can be approximated by a Continuous-time MDP in an augmented state space. The state space includes an expectation of future events, and the durations are approximated by phase-type distributions. The resulting planner is known as Tempastic-DTP [256]. More heuristic solvers have investigated the use of Generate, Test and Debug ideas by determinizing the domain to generate a policy, and debugging it by finding where it breaks down [255]. Some intermediate points have also been explored, such as incorporating duration uncertainty with deterministic action effects [16; 17].

An interesting variation on the problem arises when actions are allowed to be *interruptible*, i.e., they can be stopped before termination. This is quite useful, e.g., when an action takes an unreasonably long time, or another action finished execution and the currently running action is no longer relevant or helpful. This problem has received relatively little attention. To our knowledge, only one paper addresses this — it proposes an extension of Hybrid AO* [149].

MDPs with a large number of concurrent, durative actions continue to be a serious challenge for researchers, but one that has applications in a variety of scenarios, like traffic control, intelligent video games, and all multi-agent settings. We hope that more researcher cycles are attributed to studying this problem in the future.

7.3   RELATIONAL MDPS

The first-order extension of MDPs, called Relational MDPs or First Order MDPs, has seen an enormous popularity in the research community in the recent years. A flat MDP models the set of states directly, whereas a typical factored MDP models a state as a set of state variables. A Relational MDP, on the other hand, models the world as a set of objects, object attributes, and relationships between them. The actions are parameterized over the objects, and the transitions are described based on the properties of the argument objects. For example, a typical logistics domain expressed as a Relational MDP may involve packages, cities, trucks, planes as various objects. The relationships may include the location of a truck or a package. Actions may involve loading and unloading of the packages. This representation closely models the real world and also captures additional structure. For example, an action like *unload* has the same dynamics, irrespective of which package is being unloaded from which truck.

Relational MDPs enable an extremely compact specification of the domain and also provide additional structure to scale solution algorithms. Common ways to specify domain transitions are through Probabilistic PDDL actions [253] or relational version of DBNs [168]. The rewards are specified in two ways: a fixed reward for achieving a first-order goal property [40] or additive rewards that add up the instances of objects that satisfy a desirable property (e.g., the number of packages accurately delivered) [100; 207].

A Relational MDP with a finite number of objects can be easily converted into a factored MDP and then to a flat MDP (instantiating all domain relationships creates the set of state variables). However, this blows up the number of state variables, and hence the number of states. The last decade has seen tremendous progress in solving Relational MDPs in first-order representations directly. A significant advantage of these approaches is their ability to plan irrespective of the size of the actual problem, since the solutions are expected to work for problems with differing numbers of objects. This makes the model extremely attractive and lends hope to the scalability of MDP algorithms to real problems.

A large number of approaches have been proposed for solving Relational MDPs. An idea common to several algorithms is to directly work with a *first-order state*. A first-order state is typically described by a first-order formula. A first-order state is actually an abstract state, one which aggregates an infinite number of possible world states (in domains with differing number of objects) that satisfy the formula. For example, a first-order state may aggregate all states that have a specific package on a truck and the truck in the city where the package is to delivered.

Different researchers have studied Relational MDPs by varying two main components – the representation of the first-order solution, and the algorithm to compute the solution.

7.3.1   SOLUTION REPRESENTATIONS

Early papers on Relational MDP divide the first-order state space into disjoint partitions, where each partition is specified by a first-order formula in situation calculus [40]. They learn a value for each first-order state. These values may be stored using first-order decision trees [97; 168]. However, the algorithms typically blow up the size of the trees, since the number of generated partitions increases rapidly. As a response, researchers have generalized the decision-diagrams to First Order Decision Diagrams (FODDs) and First Order Algebraic Decision Diagrams (FOADDs) to represent these functions compactly [120; 208]. A lot of emphasis has gone into reducing the size of FODDs to improve performance.

An alternative to the previous representation is to learn values for first-order basis functions. Here, value of a state is approximated as an aggregate over the basis functions that may hold in that state. Linear combination is one common way to aggregate these basis functions [100; 205; 247], but others have also been studied, e.g., graph kernels [80]. Automatic construction of these basis functions has also received significant attention, and regression based ideas have been explored (similar to ReTrASE in Section 6.4.1) [206; 247].

An alternative formulation directly stores the first-order policy, instead of learning a value for each state. Here, the computation is over functions from first-order states to parameterized actions. These may be represented as decision lists [248].

7.3.2 ALGORITHMS

The Relational MDP algorithms can be broadly classified into two categories: deductive and inductive. The deductive algorithms use regression from the goal specification to construct the situations relevant to reaching the goal. This results in the definition of a Relational Bellman backup operator [125].

We can learn a value for a first-order state using variations of value iteration [40; 113; 125] or policy iteration [90; 239]. Approximate linear programming approaches have also been very popular, especially in conjunction with the basis function value representation [100; 205]. RTDP-like algorithms have also been designed [119].

On the other hand, inductive approaches use machine learning techniques to learn the solutions. They typically create smaller instances of a relational domain (say, with fewer number of objects). This can be grounded into a flat MDP and an exact value/policy can be learned for them. These solutions from smaller domains provide valuable training data to generalize a first-order value function or a first-order policy using machine learning [97; 168; 248].

Other approaches have also been proposed. For instance, relational extension of envelope-based planning, which starts from a seed plan, computes an envelope of states and incrementally expands fringe states, is an alternative solution technique [110].

7.4 GENERALIZED STOCHASTIC SHORTEST PATH MDPS

We have so far restricted ourselves to the SSP model from Definition 2.20. Recall that SSP MDPs make two key assumptions on the domain — (1) there exists a proper policy and (2) each improper policy π must have $V^\pi(s) = \infty$ for some state s. SSP is a relatively general model and has finite-horizon MDPs, infinite-horizon discounted MDPs as its special cases.

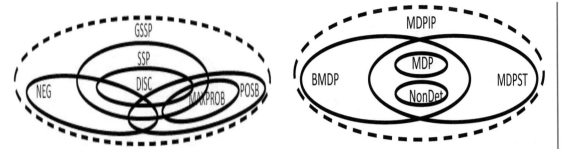

Figure 7.1: Hierarchies of MDP Models. (a) GSSPs encompass several common MDP classes (adapted from [137]). (b) MDPIPs encompass several extensions of MDPs (adapted from [235]).

Still, there are important scenarios that cannot be modeled as an SSP MDP. We have already seen two such examples. One of them is MAXPROB$_{s_0}$ MDP from Definition 6.5. *MAXPROB$_{s_0}$* is aimed at maximizing the probability of goal achievement from each state. The second is SSPADE MDP from Definition 4.19, which is an SSP-like MDP with dead ends, where all dead ends can be avoided from the known start state. There are other MDPs that are not in *SSP*. One of them models a domain with dead ends where the dead end penalty is infinite. This model is known as iSSPUDE MDPs [136]. Notice that both these models do away with the requirement of a proper policy, but are still well-formed, i.e., we can prove that the value function is bounded (at least for the states in an optimal policy rooted at s_0 when a start state is known).

Yet another class of MDP beyond the SSP model is Positive-bounded MDPs (POSB) [197]. Positive-bounded MDPs are reward-oriented MDPs with no goals. They allow arbitrary rewards, but require each state to have one positive-reward action. A second condition requires the expected value of the sum of positive rewards under each policy to be bounded. Finally, the Negative MDP model (NEG), another non-SSP type of MDP, has only negative rewards (they can be interpreted as costs), and requires the existence of a proper policy [197].

Even though *SSP* covers a broad range of important MDPs, some applications do require these non-SSP models. For example, when the objective is to avoid dead ends at all costs, MAXPROB$_{s_0}$ or iSSPUDE MDPs are natural choices. In fact, the first three international probabilistic planning competitions used domains with the MAXPROB$_{s_0}$ evaluation criterion. Positive-bounded MDPs are employed for optimal stopping problems, which are used in finance applications (e.g., to answer questions like "when to sell a house?").

Not much AI research has studied these models and their algorithms. An extension to the SSP model definition that encompasses all aforementioned models (except iSSPUDE MDPs) is known as *Generalized Stochastic Shortest Path* (GSSP) MDP [137]:

Definition 7.1   Generalized Stochastic Shortest-Path MDP. A *generalized stochastic shortest-path* (GSSP) MDP is a tuple $\langle \mathcal{S}, \mathcal{A}, \mathcal{T}, \mathcal{R}, \mathcal{G}, s_0 \rangle$ where $\mathcal{S}, \mathcal{A}, \mathcal{T} : \mathcal{S} \times \mathcal{A} \times \mathcal{S} \rightarrow [0, 1], \mathcal{G} \subseteq \mathcal{S}$ and

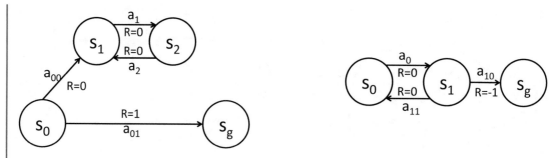

Figure 7.2: Example GSSP MDPs. (a) All states with $V^*(s) = 0$ is a fixed point, but is not an optimal fixed point. (b) The optimal value function $V^* = -1$ for both s_0 and s_1 has two greedy policies only one of which is proper.

$s_0 \in \mathcal{S}$ are as in Definition 4.3. $\mathcal{R} : \mathcal{S} \times \mathcal{A} \times \mathcal{S} \to \mathbb{R}$ is the reward obtained per state-action-state transition. A GSSP MDP satisfies two conditions:

- There exists at least one proper policy rooted at s_0,

- For all states s reachable from s_0 under any policy, and for all policies π: $V_+^\pi(s) < \infty$. $V_+^\pi(s)$ is defined as the expected sum of *non-negative* rewards when executing the policy π starting at s. In other words, starting at any reachable state, the infinite-horizon sum of non-negative rewards is bounded from above.

The objective of a GSSP MDP is to find a reward-maximizing policy that reaches a goal state with probability 1, i.e.,

$$\pi^* = \arg\max_{\pi \ proper \ rooted \ at \ s_0} V^\pi \tag{7.4}$$

Accordingly, we define $V^*(s)$ to be

$$V^* = \sup_{\pi \ proper \ rooted \ at \ s_0} V^\pi \tag{7.5}$$

Note two important subtleties in the GSSP MDP definition. First, the initial state is an integral part of a GSSP MDP instance. This class would be ill-defined without the knowledge of it. Second, when selecting an optimal policy, Equation 7.4 considers only *proper* policies, whereas *SSP's* optimal policy is selected among *all* existing policies. Why do we need to make this distinction? The reason is that in SSP MDPs, the reward-maximizing (or cost-minimizing) policy is *always* proper, whereas in a GSSP MDP this may not be so. This is best illustrated in Figure 7.2. We discuss this and other mathematical properties of GSSP MDPs in the next section.

We can prove that MAXPROB, POSB, NEG MDP models mentioned above are all special cases of *GSSP* (see Figure 7.1(a)), though the proofs are out of scope of the book.

7.4.1 MATHEMATICAL PROPERTIES

We first define a set of Bellman equations for a GSSP MDP:

$$
\begin{aligned}
V^*(s) &= 0 \qquad\qquad (\text{if } s \in \mathcal{G}) \\
&= \max_{a \in \mathcal{A}} Q^*(s, a) \quad (s \notin \mathcal{G}) \\
Q^*(s, a) &= \sum_{s' \in \mathcal{S}} \mathcal{T}(s, a, s') \big[\mathcal{R}(s, a, s') + V^*(s') \big]
\end{aligned}
\tag{7.6}
$$

GSSP MDPs have many unintuitive properties, which make their algorithms rather tricky. These include the existence of multiple fixed points (SSP MDPs have only one, corresponding to the optimal value function), and the fact that policies greedy with respect to the optimal value function may be improper.

Theorem 7.2   For a GSSP MDP, the Bellman equations (Equation 7.6) may have multiple suboptimal fixed-point value functions.

Theorem 7.3   For a GSSP MDP, not all policies greedy with respect to V^*, the optimal fixed-point value function, are proper at s_0. However, all optimal policies are greedy with respect to V^*.

Both of these theorems are easily illustrated with the example GSSPs of Figure 7.2. In the first MDP, notice that there are several fixed points, characterized by $V(s_1) = V(s_2)$ and $V(s_0) = \max\{1, V(s_1)\}$. In the second example, the optimal value function is $V^*(s_0) = V^*(s_1) = -1$. This value function has two greedy policies π_0 and π_1: $\pi_0(s_0) = a_0, \pi_0(s_1) = a_{10}$; $\pi_1(s_0) = a_0, \pi_1(s_1) = a_{11}$. Only π_0 is proper.

7.4.2 ALGORITHMS FOR GSSP MDPS

Because of these properties, the basic dynamic programming needs some changes. First, note, that the notion of the optimal value function for the complete state space may not even be finite. Thus, VI for GSSPs may not converge. Because we wish to find only a partial policy rooted at s_0, we can live with any values for dead-end states, as long as the states visited by an optimal partial policy rooted at s_0 have accurate values. We denote the subset of the state space consisting of non-dead-end states reachable from s_0 as S_p.

Second, due to multiple fixed points, VI does not necessarily converge to the optimal value function even for the states in S_p. However, the convergence is guaranteed on the subset of states in S_p if the initialization is a lower bound of the optimal, i.e., "strictly inadmissible." I.e., $\forall s \in S_p$: $\lim_{n \to \infty} V_n(s) = V^*(s)$, if $V_0(s) \le V^*(s)$.

This gives us a starting point for solving GSSP MDPs. Even though VI may not converge on all states, it does converge on the non-dead-end states. Since GSSP definition insures that there is at least one proper policy rooted at s_0, we wish to identify the subset of states in that policy and find the optimal values for those. This kind of reasoning is best fit for a heuristic search algorithm.

Notice a key challenge in designing a heuristic search algorithm for GSSPs – even on states in S_p, VI is guaranteed to converge only when $V_0 \leq V^*$; however, heuristic search succeeds when the initialization is admissible, i.e., $V_0 \geq V^*$ (for maximization problems). Thus, a heuristic search algorithm initialized admissibly will have to explicitly deal with multiple fixed points.

The details of the heuristic search approach for GSSPs, known as FRET [137], are some-what involved. The main intuition is that the basic FIND-and-REVISE algorithm (Algorithm 4.1) is insufficient for GSSP MDPs. This is because FIND-and-REVISE may not converge to an optimal value function. GSSP MDPs' conditions imply that this can happen only due to zero-reward cycles (examples are s_1-s_2 in Figure 7.2(a), and s_0-s_1 in Figure 7.2(b)). After the convergence of FIND-and-REVISE, an additional step is needed that rids the greedy policy of these zero-reward cycles (also called traps). After this trap elimination step, FIND-and-REVISE is repeated. This process continues until the greedy policy from the convergence of FIND-and-REVISE is free of all traps. The resulting algorithm is known as FIND, REVISE, ELIMINATE TRAPS (FRET) [137].

FRET, as described above, converges to the optimal value function. However, from Theorem 7.3, all greedy policies may not be optimal. This implies that the step that converts an optimal value function to an optimal policy is not as straightforward as in SSP MDPs. FRET backchains from goal states constructing a policy in a piecemeal fashion ensuring that each picked action is guaranteed to lead an agent toward the goal. For more details, please refer to the original paper [137].

GSSP generalizes *SSP* and several other MDP classes. However, it still does not capture some goal-oriented MDPs that are well-formed. For example, it does not cover iSSPUDE MDPs that have the dual optimization criterion of first selecting the policies that maximize the probability of goal achievement and then optimizing expected cost only among the selected policies [136]. Other MDPs not handled by GSSPs are ones that have zero-reward cycles but each individual reward on the cycle is not necessarily zero. Such MDPs violate the second constraint in the GSSP definition, since the sum of non-negative rewards can now be unbounded. Recently, two extensions to GSSPs have been proposed: Path-Constrained MDPs (PC-MDPs) and Stochastic Safest and Shortest Path MDPs (S^{3}P MDPs) [231; 232]. Still, our understanding with regard to each model's scope within goal-oriented MDPs is limited. Moreover, much more research is needed to design efficient algorithms for these general classes of MDPs.

7.4.3   SIXTHSENSE: A HEURISTIC FOR IDENTIFYING DEAD ENDS

How can we compute a good heuristic for a GSSP MDP? Since these MDPs have to deal with dead-end states, a good heuristic not only estimates the reward toward reaching a goal state, it also identifies states from which goals may be unreachable. The latter is also important for other MDPs with dead ends, such as fSSPUDE and iSSPUDE MDPs (discussed in Definition 4.21).

Relatively little attention has been given to identifying dead-end states for heuristic computation. We know of only one such algorithm, called SixthSense [133; 134]. The details of SixthSense are beyond the scope of this book. At a high level, SixthSense identifies dead ends in three steps. First, it determinizes the domain using the all-outcome determinization and invokes a classical planner to get sample trajectories from various start states. Using a simple timeout mechanism it also obtains a list of states from which the classical planner is unable to find a plan within the timeout. SixthSense stores such states as *potential dead ends*. Second, as it obtains more data regarding non-dead-end and potential dead-end states, it uses machine learning to find *potential nogoods*. A *nogood* is a conjunction of literals whose presence in a state guarantees that the state is a dead end. The main idea of the learner is that a potential nogood is likely to hold in many potential dead-end states and will never hold in any state that is known to be a non-dead-end. Finally, for each discovered potential nogood, SixthSense uses a planning-graph-based polynomial data structure to check whether it is truly a nogood.

Once a literal conjunction is proved to be a nogood, it is added to the set of known nogoods. All states where a nogood holds are proved to be dead ends, and their values are initialized according to dead-end penalty for SSPUDE MDPs, or $-\infty$ for GSSPs, or 0 for MAXPROB$_{s_0}$ MDPs. SixthSense may not always be able to prove that a potential nogood is a true nogood using its polynomial algorithm; hence, it is a sound but not complete method to identify dead-end states. The use of SixthSense has been repeatedly shown to be valuable for domains with large numbers of dead ends [133; 134; 136; 137].

7.5 OTHER MODELS

There are still a large number of extensions that we are unable to discuss in this book due to space constraints. Some notable ones are briefly mentioned below:

1. **Non-deterministic Planning:** This formalism (briefly introduced in Section 6.6) models action uncertainty by allowing an action to have a set of potential successors, but the probabilities of the transitions are not modeled and, commonly, action costs are ignored [54]. Non-deterministic planning studies satisficing solutions, rather than the cost-optimized ones (which are studied in MDP theory). It attempts to find three kinds of solutions: *weak plan*, a plan that may reach the goal, *strong plan*, a plan that is guaranteed to reach the goal, and *strong-cyclic plan*, a plan which may loop, but as long as nature is fair in generating the next state, it will eventually reach the goal. A proper policy for an MDP is equivalent to a strong-cyclic plan in the non-deterministic version of the domain.

2. **Multi-Objective MDPs:** Often a problem has several competing objectives, e.g., total time, monetary cost, battery usage, etc. It is possible to convert a problem with multiple objectives into a standard MDP by defining a single utility function that is a complex combination of all such objectives. Typically, however, it is difficult to construct such a utility function *a priori*. In response, researchers have studied Multi-Objective MDPs or Multi-Criteria MDPs [52; 238], which jointly optimize all these metrics. They return a pareto-set of all non-dominated policies (a policy may not dominate the other since it may be better for one objective but worse for the other).

3. **MDPs with Imprecise/Uncertain Probabilities:** MDPIPs are an extension of the MDP model where instead of a known probability distribution, a set of probability distributions is provided for each action [117; 213]. MDPIPs study several optimization criteria, like finding the best policy for the worst-case transition probabilities (pessimistic view), or for best case transition probabilities (optimistic view), or finding policies that are best for at least one probability distribution. Several MDP algorithms (dynamic programming-based, LP-based) and optimizations (e.g., ADD extensions) have been extended to MDPIPs [77].

MDPIPs are a general model, and have many interesting subcases. One of them is *MDPs with Set-Valued Transitions*. MDPSTs explore the continuum between non-deterministic planning and MDPs, and unify both the models. They associate a probability with a set of states. Common MDP algorithms have been extended to handle MDPSTs, e.g., labeled RTDP [235].

Another special case of MDPIPs is *Bounded-Parameter MDPs* [94]. BMDP models MDPs where a range of probability values is known for each transition. RTDP has been extended to BMDPs. Figure 7.1(b) (adapted from [235]) shows the inter-relationships between the several models.

4. **Reinforcement Learning:** RL studies the scenarios in which the agent doesn't have access to the complete transition (and/or the reward model). It has to act to discover the model while maximizing the reward. RL is a strictly harder problem than planning with MDPs. RL is a large area of research within AI, perhaps even larger than probabilistic planning. The model is quite popular in the robotics community, because often the exact transition model is unknown, but one has access to the actual robot/simulator. Alternately, occasionally, the exact reward is hard to specify, but the behavior is known, e.g., if the objective is to drive safely. For an extensive discussion on RL please refer to the several books written on the subject (e.g., [224; 226]).

In the last few years, a special case of RL by the name of *Monte Carlo Planning* has become popular. Here, it is assumed that the algorithm has access to a domain simulator. One could simulate an agent's actions on the simulator to learn about the domain. This can be quite beneficial in several scenarios like when the domains are too complex to specify in a planning language, e.g., modeling atmospheric processes, migration of birds, etc. The approaches for Monte Carlo Planning commonly use versions and enhancements of the UCT algorithm discussed in Section 6.2.1. Results have shown tremendous success in building AI agents playing Solitaire and real-time strategy games using Monte Carlo Planning [10; 26].

5. **Partially Observable Markov Decision Process (POMDP):** As the name suggests, POMDPs model scenarios where the agent cannot observe the world state fully [123]. A POMDP agent needs to execute actions for two reasons: for changing the world state (as in an MDP) and for obtaining additional information about the current world state. As Section 7.1.1 explains, a POMDP is a large Continuous MDP, in which a state-variable is the world state, and its value denotes the agent's belief (probability) that it is in that state. Straightforward

implementations of MDP algorithms do not scale up to POMDPs and, over the years, a large number of specialized POMDP techniques have been developed, with successes in scaling the algorithms to millions of states [214]. POMDPs have also seen several applications, e.g., dialog management [241], intelligent control of workflows [65], intelligent tutoring [200], and several robotic planning applications [233].

6. **Collaborative Multi-agent Models**: Several extensions have been proposed to model domains with multiple agents sharing a common goal. One of the first is Multi-Agent MDPs [35], where the full state is visible to all, but additional coordination mechanisms are needed to achieve best collaboration. Decentralized MDPs are an extension where the full state is not visible to every agent, but each part of the state space is visible to some agents [20]. Decentralized POMDPs (Dec-POMDP) are a concurrent extension to POMDPs, in which some part of the state space may not be observed by any agent [190]. In all these models, policy computation typically happens in a centralized manner. A closely related model is Interactive POMDP [95], which also extends POMDPs to multi-agent settings. Interactive POMDPs explicitly reason about the beliefs of other agents. This leads to infinite nesting, although this belief asymptotes.

7. **Adversarial Multi-agent Models:** Markov Games extend MDPs to self-interested agents [155]. Here different agents have individual rewards. In some cases, the rewards may be zero-sum leading to an adversarial game, but that is not a requirement of the model. Partially Observable Stochastic Games (POSGs) [104] extend this scenario to partially observable settings.

7.6   ISSUES IN PROBABILISTIC PLANNING

Before we conclude the book, we briefly discuss some important non-technical issues concerning the probabilistic planning community.

7.6.1   THE IMPORTANCE OF PLANNING COMPETITIONS

One of major forces that drives research forward in the planning community is the International Planning Competition (IPC).[3] The IPC was created in 1998 and since then it has been held (almost) every two years. It is overseen by the ICAPS Council; ICAPS, the International Conference on Automated Planning and Scheduling, is a premier conference for AI planning researchers and practitioners.

IPC has set a common ground for various planning researchers to compete on a common set of benchmark problems. It helps assess and understand the relative merits of different research ideas and spurs the progress of the field in seemingly more rewarding directions. Recent competitions have required the competitors to provide release software that further enhances the replicability of results and rapid innovations.

[3]http://ipc.icaps-conference.org/

Starting from 2004, IPC has also included a probabilistic track. The international *probabilistic* planning competitions (IPPC) have brought in probabilistic planners, much in the same way as IPC did for the classical planners. The first three competitions tested the solvers on goal-oriented domains that have possibly catastrophic effects leading to dead ends. The focus was less on cost-optimization and more on goal reachability. The domain representation used was Probabilistic PDDL. Many of the benchmarks domains were probabilistic extensions of popular classical planning domains such as blocksworld. The fourth IPPC (2011) changed IPPC's course by its use of RDDL as the description language, and its emphasis on MDPs with huge branching factors, and reward maximization problems.

Planning competitions have been quite controversial for different reasons, but have also had tremendous influence on the research directions of the community. For probabilistic planning it is easy to follow the research trajectories in the context of the challenges posed by the competition. The whole sub-field of determinization-based approximations got developed due to the success of FF-Replan in 2004 and 2006. FF-Replan was the winner of IPPC 2004 and the unofficial winner of IPPC 2006, since none of the competing solvers could outperform it. The success of such a simple solver was considered quite controversial and raised significant questions on the choice of domains in IPPC. At the same time, this also encouraged researchers to investigate determinization-based ideas. This led to the development of RFF and HMDPP, which comprehensively defeated FF-Replan in IPPC 2008.

IPC 2011 represented domains in a DBN-like representation, and the domains problems were reward-oriented. The rationale was that this representation allows for modeling more realistic problems such as traffic management and computer lab administration. This representation makes it harder to apply domain-determinization ideas, and researchers have investigated sampling based approaches (e.g., UCT) to deal with the large sizes of the domains. Still, a sudden and substantial change of representation and domain characteristics makes IPPC 2011 somewhat controversial too.

So far, the IPPCs have seen a relatively thin participation. Where IPC 2011 saw 27 participants in its flagship classical planning sequential satisficing track, IPPC only got 5 competitors in the MDP track. We hope that slowly the competition will grow and more researchers from different sub-disciplines will implement MDP planners and compete on common ground.

7.6.2   THE BANE OF MANY CONFERENCES

An interesting characteristic of the MDP sub-community within AI is the availability of several venues interested in this research. MDP papers are frequently published in the planning conference (ICAPS), uncertainty reasoning conference (UAI), machine learning conferences (e.g., ICML), neural systems conference (NIPS), robotics conferences (e.g., ICRA), and the umbrella AI conferences – AAAI, ICJAI, and others. At one level, this is a reflection of the generality of the MDP model and the diverse value it brings to the broader AI theme. On the other hand, however, this is not the best situation for keeping abreast of the recent advances in MDP literature. A dedicated venue for

MDP researchers will allow an easier transmission of ideas and enable rapid progress in this field of inquiry.

A related issue is the flow of ideas among other disciplines of science and technology that use MDPs. As discussed in Chapter 1, MDPs are regularly employed in operations research, control theory, computational finance, and other disciplines. While the techniques get specialized for the specific use cases, a distillation of the important general ideas and their dissemination to different communities would be especially valuable. We believe that there is a need to create such a venue where all MDP researchers get together and learn from each other's experience.

7.7 SUMMARY

Markov Decision Processes are a fundamental formalism for modeling sequential decision making problems in stochastic domains. They enjoy a wide popularity in the AI literature – they are integral in designing an intelligent, rational agent, capable of sensing the uncertain world and acting for long periods of time in order to achieve its objective. They are applicable to a large number of applications, including game playing, robotics, military operations planning, intelligent tutoring, and design of intelligent games.

Although they are a simple and expressive model, the basic set of solution techniques do not quite scale to many real world problems, and the key challenge for the researchers has been algorithm scalability. In this book, we have surveyed the multitude of ideas that go into making MDP algorithms efficient. The fundamental solution algorithms are based on iterative dynamic programming, although a linear programming approach is also possible. Modern optimal algorithms draw from a vast repertoire of techniques, like graph algorithms, heuristic search, compact value function representations, and simulation-based approaches.

Of course, optimal solutions for such an expressive model are a luxury. An enormous number of approximation algorithms have been suggested that exploit several intuitions, such as inadmissible heuristics, interleaving planning and execution, special processing for dead-end states, domain determinization ideas, hybridizing multiple algorithms, hierarchical problem decompositions, and so on. These algorithms explore the vast space of optimality-efficiency tradeoffs.

Finally, we have briefly described a few important extensions to the expressiveness of the basic MDP model. Substantial literature developing each one of these extensions is available. It is impossible to do justice to these, since one could write a similar-sized survey on almost each advanced topic. Our hope is that our brief writeups give a flavor of the key ideas and a starting point to an interested reader.

It is important to note that in this survey we have focused exclusively on *domain-independent planning*, i.e., ideas that are applicable irrespective of the domain at hand. These develop our understanding of the fundamental properties and solution techniques for the model. Even though almost all decision making problems may be cast either as an MDP or as one of its extensions, solving them using the domain-independent techniques is often not feasible. In practice, additional domain-specific analysis is performed to model the particular structure of the domain that could make computation more efficient.

Last but not the least, we stress the relative lack of software suites for solving MDPs. Besides a handful of software packages that implement specific algorithms, not much is available for an application designer to use. Given the amount of research that has already been done in the area, the time is ripe for releasing some of the important algorithms in an optimized code-base. A Matlab-like software for MDPs may go a long way in taking MDPs (and extensions) from research to practice, since it will allow users, who may not understand the theory, to test the techniques on their problems. A standardized code-base is likely to speed up progress in the research community as well.

Bibliography

[1] Douglas Aberdeen and Olivier Buffet. Concurrent probabilistic temporal planning with policy-gradients. In *Proceedings of the Fifth International Conference on Automated Planning and Scheduling*, pages 10–17, 2007. Cited on page(s) 150

[2] Douglas Aberdeen, Sylvie Thiébaux, and Lin Zhang. Decision-theoretic military operations planning. In *Proceedings of the Second International Conference on Automated Planning and Scheduling*, pages 402–412, 2004. Cited on page(s) 1, 150

[3] Charles J. Alpert. Multi-way graph and hypergraph partitioning. Ph.D. thesis, University of California, Los Angeles, 1996. Cited on page(s) 51

[4] David Andre, Nir Friedman, and Ronald Parr. Generalized prioritized sweeping. In *Advances in Neural Information Processing Systems*, 1997. Cited on page(s) 46

[5] David Andre and Stuart J. Russell. State abstraction for programmable reinforcement learning agents. In *Proceedings of the Eighteenth National Conference on Artificial Intelligence*, pages 119–125, 2002. Cited on page(s) 134

[6] W. Ross Ashby. *Design for a Brain*. Chapman and Hall, 1952. Cited on page(s) 135

[7] W. Ross Ashby. *An Introduction to Cybernetics*. Chapman and Hall, 1956. Cited on page(s) 135

[8] Fahiem Bacchus and Froduald Kabanza. Using temporal logics to express search control knowledge for planning. *Artificial Intelligence*, 116(1-2):123–191, 2000. DOI: 10.1016/S0004-3702(99)00071-5 Cited on page(s) 136

[9] R. Iris Bahar, Erica A. Frohm, Charles M. Gaona, Gary D. Hachtel, Enrico Macii, Abelardo Pardo, and Fabio Somenzi. Algebraic decision diagrams and their applications. In *Proceedings of the IEEE/ACM International Conference on Computer-Aided Design*, pages 188–191, 1993. DOI: 10.1109/ICCAD.1993.580054 Cited on page(s) 83, 87

[10] Radha-Krishna Balla and Alan Fern. UCT for tactical assault planning in real-time strategy games. In *Proceedings of the Twenty-first International Joint Conference on Artificial Intelligence*, pages 40–45, 2009. Cited on page(s) 158

[11] Jennifer L. Barry, Leslie Pack Kaelbling, and Tomás Lozano-Pérez. Deth*: Approximate hierarchical solution of large Markov decision processes. In *Proceedings of the Twenty-second*

International Joint Conference on Artificial Intelligence, pages 1928–1935, 2011. DOI: 10.5591/978-1-57735-516-8/IJCAI11-323 Cited on page(s) 137

[12] A. Barto, S. Bradtke, and S. Singh. Learning to act using real-time dynamic programming. *Artificial Intelligence*, 72:81–138, 1995. DOI: 10.1016/0004-3702(94)00011-O Cited on page(s) 69

[13] Andrew G. Barto and Sridhar Mahadevan. Recent advances in hierarchical reinforcement learning. *Discrete Event Dynamic Systems*, 13(4):341–379, 2003. DOI: 10.1023/A:1022140919877 Cited on page(s) 127

[14] Nicole Bauerle and Ulrich Rieder. *Markov Decision Processes with Applications to Finance*. Springer, 2011. DOI: 10.1007/978-3-642-18324-9 Cited on page(s) 3

[15] Jonathan Baxter, Peter L. Bartlett, and Lex Weaver. Experiments with infinite-horizon, policy-gradient estimation. *Journal of Artificial Intelligence Research*, 15:351–?381, 2001. Cited on page(s) 126

[16] Eric Beaudry, Froduald Kabanza, and François Michaud. Planning for concurrent action executions under action duration uncertainty using dynamically generated bayesian networks. In *Proceedings of the Eighth International Conference on Automated Planning and Scheduling*, pages 10–17, 2010. Cited on page(s) 150

[17] Eric Beaudry, Froduald Kabanza, and François Michaud. Planning with concurrency under resources and time uncertainty. In *Proceedings of the Nineteenth European Conference on Artificial Intelligence*, pages 217–222, 2010. DOI: 10.3233/978-1-60750-606-5-217 Cited on page(s) 150

[18] Richard Bellman. *Dynamic Programming*. Prentice Hall, 1957. Cited on page(s) 1, 38

[19] Abder Rezak Benaskeur, Froduald Kabanza, Eric Beaudry, and Mathieu Beaudoin. A probabilistic planner for the combat power management problem. In *Proceedings of the Sixth International Conference on Automated Planning and Scheduling*, pages 12–19, 2008. Cited on page(s) 150

[20] Daniel S. Bernstein, Robert Givan, Neil Immerman, and Shlomo Zilberstein. The complexity of decentralized control of Markov decision processes. *Mathematics of Operations Research*, 27(4):819–840, 2002. DOI: 10.1287/moor.27.4.819.297 Cited on page(s) 159

[21] Dimitri P. Bertsekas. *Dynamic Programming and Optimal Control*. Athena Scientific, 1995. Cited on page(s) 4, 20, 21, 37, 38, 43, 51

[22] Dimitri P. Bertsekas. *Dynamic Programming and Optimal Control*, volume 2. Athena Scientific, 2000. Cited on page(s) 75

[23] Dimitri P. Bertsekas and John N. Tsitsiklis. *Parallel and Distributed Computation: Numerical Methods.* Prentice-Hall, 1989. Cited on page(s) 44

[24] Dimitri P. Bertsekas and John N. Tsitsiklis. *Neuro-Dynamic Programming.* Athena Scientific, 1996. Cited on page(s) 4, 22, 36, 43

[25] Venkata Deepti Kiran Bhuma and Judy Goldsmith. Bidirectional LAO* algorithm. In *Proceedings of the Eighteenth International Joint Conference on Artificial Intelligence*, pages 980–992, 2003. Cited on page(s) 67

[26] Ronald Bjarnason, Alan Fern, and Prasad Tadepalli. Lower bounding Klondike Solitaire with Monte-Carlo planning. In *Proceedings of the Seventh International Conference on Automated Planning and Scheduling*, 2009. Cited on page(s) 158

[27] Avrim Blum and John Langford. Probabilistic planning in the graphplan framework. In *Proceedings of the Fifth European Conference on Planning*, pages 319–332, 1999. DOI: 10.1007/10720246_25 Cited on page(s) 149

[28] Beate Bollig and Ingo Wegener. Improving the variable ordering of OBDDs is NP-complete. *IEEE Transactions on Computers*, 45(9):993–1002, 1996. DOI: 10.1109/12.537122 Cited on page(s) 87

[29] Blai Bonet. On the speed of convergence of value iteration on stochastic shortest-path problems. *Mathematics of Operations Research*, 32(2):365–373, 2007. DOI: 10.1287/moor.1060.0238 Cited on page(s) 36, 43

[30] Blai Bonet and Hector Geffner. Planning as heuristic search. *Artificial Intelligence*, 129:5–33, 2001. DOI: 10.1016/S0004-3702(01)00108-4 Cited on page(s) 79, 115

[31] Blai Bonet and Hector Geffner. Faster heuristic search algorithms for planning with uncertainty and full feedback. In *Proceedings of the Eighteenth International Joint Conference on Artificial Intelligence*, pages 1233–1238, 2003. Cited on page(s) 61, 64

[32] Blai Bonet and Hector Geffner. Labeled RTDP: Improving the convergence of real-time dynamic programming. In *Proceedings of the First International Conference on Automated Planning and Scheduling*, pages 12–21, 2003. Cited on page(s) 70, 71

[33] Blai Bonet and Hector Geffner. Learning depth-first search: A unified approach to heuristic search in deterministic and non-deterministic settings, and its application to MDPs. In *Proceedings of the Fourth International Conference on Automated Planning and Scheduling*, pages 3–23, 2006. Cited on page(s) 64

[34] Blai Bonet, Gábor Lorincs, and Hector Geffner. A robust and fast action selection mechanism for planning. In *Proceedings of the Fourteenth National Conference on Artificial Intelligence*, pages 714–719, 1997. Cited on page(s) 78

[35] Craig Boutilier. Sequential optimality and coordination in multiagent systems. In *Proceedings of the Sixteenth International Joint Conference on Artificial Intelligence*, pages 478–485, 1999. Cited on page(s) 159

[36] Craig Boutilier, Thomas Dean, and Steve Hanks. Decision theoretic planning: Structural assumptions and computational leverage. *Journal of Artificial Intelligence Research*, 11:1–94, 1999. DOI: 10.1613/jair.575 Cited on page(s) 1

[37] Craig Boutilier and Richard Dearden. Approximate value trees in structured dynamic programming. In *Proceedings of the Thirteenth International Conference on Machine Learning*, pages 54–62, 1996. Cited on page(s) 95

[38] Craig Boutilier, Richard Dearden, and Moises Goldszmidt. Exploiting structure in policy construction. In *Proceedings of the Fourteenth International Joint Conference on Artificial Intelligence*, pages 1104–1113, 1995. Cited on page(s) 91

[39] Craig Boutilier, Richard Dearden, and Moises Goldszmidt. Stochastic dynamic programming with factored representations. *Artificial Intelligence*, 121(1-2):49–107, 2000. DOI: 10.1016/S0004-3702(00)00033-3 Cited on page(s) 91, 95

[40] Craig Boutilier, Ray Reiter, and Bob Price. Symbolic dynamic programming for first-order MDPs. In *Proceedings of the Seventeenth International Joint Conference on Artificial Intelligence*, pages 690–697, 2001. Cited on page(s) 151, 152

[41] Justin A. Boyan and Michael L. Littman. Exact solutions to time-dependent MDPs. In *Advances in Neural Information Processing Systems*, pages 1026–1032, 2000. DOI: 10.1103/PhysRevA.26.729 Cited on page(s) 143, 145, 148

[42] Steven J. Bradtke and Michael O. Duff. Reinforcement learning methods for continuous-time Markov decision problems. In *Advances in Neural Information Processing Systems*, pages 393–400, 1994. Cited on page(s) 128, 148

[43] John L. Bresina, Richard Dearden, Nicolas Meuleau, Sailesh Ramkrishnan, David E. Smith, and Richard Washington. Planning under continuous time and resource uncertainty: A challenge for ai. In *Proceedings of the Eighteenth Conference on Uncertainty in Artificial Intelligence*, pages 77–84, 2002. Cited on page(s) 143, 148

[44] Randal E. Bryant. Graph-based algorithms for boolean function manipulation. *IEEE Transactions on Computers*, 35(8):677–691, 1986. Cited on page(s) 83

[45] Randal E. Bryant. Symbolic boolean manipulation with ordered binary-decision diagrams. *ACM Computing Surveys*, 24(3):293–318, 1992. DOI: 10.1145/136035.136043 Cited on page(s) 87

[46] Daniel Bryce and Olivier Buffet.   International planning competition, uncertainty part: Benchmarks and results. In http://ippc-2008.loria.fr/wiki/images/0/03/ Results.pdf, 2008. Cited on page(s) 108

[47] Daniel Bryce and Seungchan Kim. Planning for gene regulatory network intervention. In *Proceedings of the Twentieth International Joint Conference on Artificial Intelligence*, pages 1834–1839, 2007. DOI: 10.1109/LSSA.2006.250382 Cited on page(s) 1

[48] Olivier Buffet and Douglas Aberdeen. The factored policy gradient planner (IPC?06 version). In *Fifth International Planning Competition (IPC-5)*, 2006. DOI: 10.1016/j.artint.2008.11.008 Cited on page(s) 126

[49] Olivier Buffet and Douglas Aberdeen. FF+FPG: Guiding a policy-gradient planner. In *Proceedings of the Fifth International Conference on Automated Planning and Scheduling*, 2006. Cited on page(s) 126, 142

[50] Olivier Buffet and Douglas Aberdeen. The factored policy-gradient planner. *Artificial Intelligence*, 173(5-6):722–747, 2009. DOI: 10.1016/j.artint.2008.11.008 Cited on page(s) 124, 126, 150

[51] Tom Bylander. The computational complexity of propositional STRIPS planning. *Artificial Intelligence*, 69 (1–2):164–204, 1994. DOI: 10.1016/0004-3702(94)90081-7 Cited on page(s) 78, 116

[52] Krishnendu Chatterjee, Rupak Majumdar, and Thomas A. Henzinger. Markov decision processes with multiple objectives. In *Proceedings of Twenty-third Annual Symposium on Theoretical Aspects of Computer Science*, pages 325–336, 2006. DOI: 10.1007/11672142_26 Cited on page(s) 157

[53] Chef-Seng Chow and John N. Tsitsiklis. An optimal one-way multigrid algorithm for discrete-time stochastic control. *IEEE Transactions on Automatic Control*, AC-36(8):898–914, 1991. DOI: 10.1109/9.133184 Cited on page(s) 144

[54] Alessandro Cimatti, Marco Pistore, Marco Roveri, and Paolo Traverso. Weak, strong, and strong cyclic planning via symbolic model checking. *Artificial Intelligence*, 147(1-2):35–84, 2003. DOI: 10.1016/S0004-3702(02)00374-0 Cited on page(s) 140, 157

[55] Thomas H. Cormen, Charles E. Leiserson, Ronald L. Rivest, and Clifford Stein. *Introduction to Algorithms, 3rd Edition*. MIT Press, 2001. Cited on page(s) 41, 51, 145

[56] Elva Corona-Xelhuantzi, Eduardo F. Morales, and L. Enrique Sucar. Solving policy conflicts in concurrent Markov decision processes. In *Proceedings of ICAPS Workshop on Planning under Uncertainty*, 2010. Cited on page(s) 149

[57] Robert H. Crites and Andrew G. Barto. Improving elevator performance using reinforcement learning. In *Advances in Neural Information Processing Systems*, pages 1017–1023, 1995. Cited on page(s) 1

[58] Peng Dai and Judy Goldsmith. LAO*, RLAO*, or BLAO*. In *Proceedings of AAAI workshop on heuristic search*, pages 59–64, 2006. Cited on page(s) 67

[59] Peng Dai and Judy Goldsmith. Topological value iteration algorithm for Markov decision processes. In *Proceedings of the Twentieth International Joint Conference on Artificial Intelligence*, pages 1860–1865, 2007. Cited on page(s) 51

[60] Peng Dai and Eric A. Hansen. Prioritizing Bellman backups without a priority queue. In *Proceedings of the Fifth International Conference on Automated Planning and Scheduling*, pages 113–119, 2007. Cited on page(s) 47, 48

[61] Peng Dai, Mausam, and Daniel S. Weld. Partitioned external memory value iteration. In *Proceedings of the Twenty-third AAAI Conference on Artificial Intelligence*, pages 898–904, 2008. Cited on page(s) 52, 53

[62] Peng Dai, Mausam, and Daniel S. Weld. Domain-independent, automatic partitioning for probabilistic planning. In *Proceedings of the Twenty-first International Joint Conference on Artificial Intelligence*, pages 1677–1683, 2009. Cited on page(s) 52

[63] Peng Dai, Mausam, and Daniel S. Weld. Focused topological value iteration. In *Proceedings of the Seventh International Conference on Automated Planning and Scheduling*, 2009. Cited on page(s) 75, 76

[64] Peng Dai, Mausam, and Daniel S. Weld. Decision-theoretic control of crowd-sourced work-flows. In *Proceedings of the Twenty-fourth AAAI Conference on Artificial Intelligence*, 2010. Cited on page(s) 145

[65] Peng Dai, Mausam, and Daniel S. Weld. Artificial intelligence for artificial artificial intelligence. In *Proceedings of the Twenty-fifth AAAI Conference on Artificial Intelligence*, 2011. Cited on page(s) 145, 159

[66] Peng Dai, Mausam, Daniel S. Weld, and Judy Goldsmith. Topological value iteration algorithms. *Journal of Artificial Intelligence Research*, 42:181–209, 2011. Cited on page(s) 51, 76

[67] Peter Dayan and Geoffrey E. Hinton. Feudal reinforcement learning. In *Advances in Neural Information Processing Systems*, pages 271–278, 1992. Cited on page(s) 135

[68] Ma. de G. Garcia-Hernandez, J. Ruiz-Pinales, A. Reyes-Ballesteros, E. Onaindia, J. Gabriel Avina-Cervantes, and S. Ledesma. Acceleration of association-rule based markov decision processes. *Journal of Applied Research and Technology*, 7(3), 2009. Cited on page(s) 55

[69] Thomas Dean and Robert Givan. Model minimization in Markov decision processes. In *Proceedings of the Fourteenth National Conference on Artificial Intelligence*, pages 106–111, 1997. Cited on page(s) 51, 137

[70] Thomas Dean, Leslie Pack Kaelbling, Jak Kirman, and Ann E. Nicholson. Planning with deadlines in stochastic domains. In *Proceedings of the Eleventh National Conference on Artificial Intelligence*, pages 574–579, 1993. Cited on page(s) 1

[71] Thomas Dean and Keiji Kanazawa. A model for reasoning about persistence and causation. *Computational Intelligence*, 5(3):142–150, 1989. DOI: 10.1111/j.1467-8640.1989.tb00324.x Cited on page(s) 26, 89

[72] Thomas Dean and Shieu-Hong Lin. Decomposition techniques for planning in stochastic domains. In *Proceedings of the Fourteenth International Joint Conference on Artificial Intelligence*, pages 1121–1129, 1995. Cited on page(s) 132, 135

[73] Richard Dearden. Structured prioritised sweeping. In *Proceedings of the Eighteenth International Conference on Machine Learning*, pages 82–89, 2001. Cited on page(s) 92

[74] Richard Dearden and Craig Boutilier. Integrating planning and execution in stochastic domains. In *Proceedings of the Tenth Conference on Uncertainty in Artificial Intelligence*, pages 162–169, 1994. Cited on page(s) 1

[75] Richard Dearden, Nicolas Meuleau, Sailesh Ramakrishman, David Smith, and Rich Washington. Incremental contingency planning. In *Proceedings of the Workshop on Planning under Uncertainty and Incomplete Information at ICAPS'03*, pages 38–?47, 2003. Cited on page(s) 106

[76] Karina Vadivia Delgado, Cheng Fang, Scott Sanner, and Leliane de Barros. Symbolic bounded real-time dynamic programming. In *Advances in Artificial Intelligence SBIA 2010*, volume 6404 of *Lecture Notes in Computer Science*, pages 193–202. 2011. DOI: 10.1007/978-3-642-16138-4_20 Cited on page(s) 92

[77] Karina Valdivia Delgado, Scott Sanner, and Leliane Nunes de Barros. Efficient solutions to factored MDPs with imprecise transition probabilities. *Artificial Intelligence*, 175(9-10):1498–1527, 2011. DOI: 10.1016/j.artint.2011.01.001 Cited on page(s) 158

[78] F. D'Epenoux. A probabilistic production and inventory problem. *Management Science*, 10:98–108, 1963. DOI: 10.1287/mnsc.10.1.98 Cited on page(s) 53

[79] Thomas G. Dietterich. Hierarchical reinforcement learning with the MAXQ value function decomposition. *Journal of Artificial Intelligence Research*, 13:227–303, 2000. DOI: 10.1613/jair.639 Cited on page(s) 130, 131, 132, 136

[80] Kurt Driessens, Jan Ramon, and Thomas Gärtner. Graph kernels and gaussian processes for relational reinforcement learning. *Machine Learning*, 64(1-3):91–119, 2006. DOI: 10.1007/s10994-006-8258-y Cited on page(s) 152

[81] Lester E. Dubins and Leonard J. Savage. *Inequalities for Stochastic Pro-cesses (How to Gamble If You Must)*. Dover Publications, 1976. Cited on page(s) 3

[82] Stefan Edelkamp, Shahid Jabbar, and Blai Bonet. External memory value iteration. In *Proceedings of the Fifth International Conference on Automated Planning and Scheduling*, pages 128–135, 2007. Cited on page(s) 53

[83] E.A. Feinberg and A. Shwartz, editors. *Handbook of Markov Decision Processes - Methods and Applications*. Kluwer International Series, 2002. DOI: 10.1007/978-1-4615-0805-2 Cited on page(s) 1

[84] Zhengzhu Feng, Richard Dearden, Nicolas Meuleau, and Rich Washington. Dynamic programming for structured continuous Markov decision problems. In *Proceedings of the Twentieth Conference on Uncertainty in Artificial Intelligence*, pages 154–161, 2004. Cited on page(s) 145

[85] Zhengzhu Feng and Eric A. Hansen. Symbolic heuristic search for factored Markov decision processes. In *Proceedings of the Eighteenth National Conference on Artificial Intelligence*, pages 455–460, 2002. Cited on page(s) 90

[86] Zhengzhu Feng, Eric A. Hansen, and Shlomo Zilberstein. Symbolic generalization for on-line planning. In *Proceedings of the Nineteenth Conference on Uncertainty in Artificial Intelligence*, pages 209–216, 2003. Cited on page(s) 92

[87] David I. Ferguson and Anthony Stentz. Focussed dynamic programming: Extensive comparative results. Technical Report CMU-RI-TR-04-13, Carnegie Mellon University, 2004. Cited on page(s) 48

[88] David I. Ferguson and Anthony Stentz. Focussed propagation of MDPs for path planning. In *Proceedings of the Sixteenth IEEE International Conference on Tools with Artificial Intelligence*, pages 310–317, 2004. DOI: 10.1109/ICTAI.2004.64 Cited on page(s) 48

[89] Alan Fern. Monte-carlo planning: Basic principles and recent progress. Tutorial at ICAPS'10, 2010. Cited on page(s) 111, 113

[90] Alan Fern, Sung Wook Yoon, and Robert Givan. Approximate policy iteration with a policy language bias. In *Advances in Neural Information Processing Systems*, 2003. Cited on page(s) 152

[91] Richard Fikes and Nils Nilsson. STRIPS: a new approach to the application of theorem proving to problem solving. *Artificial Intelligence*, 2:189–208, 1971. DOI: 10.1016/0004-3702(71)90010-5 Cited on page(s) 79

[92] Janae N. Foss and Nilufer Onder. A hill-climbing approach for planning with temporal uncertainty. In *Proceedings of the Nineteenth International Florida Artificial Intelligence Research Society Conference*, 2006. Cited on page(s) 148

[93] Sylvain Gelly and David Silver. Achieving master level play in 9x9 computer Go. In *Proceedings of the Twenty-third AAAI Conference on Artificial Intelligence*, pages 1537–1540, 2008. Cited on page(s) 113

[94] Robert Givan, Sonia M. Leach, and Thomas Dean. Bounded-parameter Markov decision processes. *Artificial Intelligence*, 122(1-2):71–109, 2000. DOI: 10.1016/S0004-3702(00)00047-3 Cited on page(s) 158

[95] Piotr J. Gmytrasiewicz and Prashant Doshi. A framework for sequential planning in multi-agent settings. *Journal of Artificial Intelligence Research*, 24:49–79, 2005. Cited on page(s) 159

[96] Judy Goldsmith, Michael L. Littman, and Martin Mundhenk. The complexity of plan existence and evaluation in probabilistic domains. In *Proceedings of the Thirteenth Conference on Uncertainty in Artificial Intelligence*, 1997. Cited on page(s) 28, 29

[97] Charles Gretton and Sylvie Thiébaux. Exploiting first-order regression in inductive policy selection. In *Proceedings of the Twentieth Conference on Uncertainty in Artificial Intelligence*, pages 217–225, 2004. Cited on page(s) 151, 152

[98] Carlos Guestrin and Geoffrey J. Gordon. Distributed planning in hierarchical factored MDPs. In *Proceedings of the Eighteenth Conference on Uncertainty in Artificial Intelligence*, pages 197–206, 2002. Cited on page(s) 136, 149

[99] Carlos Guestrin, Milos Hauskrecht, and Branislav Kveton. Solving factored MDPs with continuous and discrete variables. In *Proceedings of the Twentieth Conference on Uncertainty in Artificial Intelligence*, pages 235–242, 2004. Cited on page(s) 146

[100] Carlos Guestrin, Daphne Koller, Chris Gearhart, and Neal Kanodia. Generalizing plans to new environments in relational MDPs. In *Proceedings of the Eighteenth International Joint Conference on Artificial Intelligence*, pages 1003–1010, 2003. Cited on page(s) 151, 152

[101] Carlos Guestrin, Daphne Koller, and Ronald Parr. Max-norm projections for factored MDPs. In *Proceedings of the Seventeenth International Joint Conference on Artificial Intelligence*, pages 673–682, 2001. Cited on page(s) 149

[102] Carlos Guestrin, Daphne Koller, Ronald Parr, and Shobha Venkataraman. Efficient solution algorithms for finite MDPs. *Journal of Artificial Intelligence Research*, 19:399–468, 2003. Cited on page(s) 122, 123, 124

[103] Eric A. Hansen. Suboptimality bounds for stochastic shortest path problems. In *Proceedings of the Twenty-seventh Conference on Uncertainty in Artificial Intelligence*, pages 301–310, 2011. Cited on page(s) 36, 43

[104] Eric A. Hansen, Daniel S. Bernstein, and Shlomo Zilberstein. Dynamic programming for partially observable stochastic games. In *Proceedings of the Nineteenth National Conference on Artificial Intelligence*, pages 709–715, 2004. Cited on page(s) 159

[105] Eric A. Hansen and Shlomo Zilberstein. LAO*: A heuristic search algorithm that finds solutions with loops. *Artificial Intelligence*, 129:35–62, 2001.
DOI: 10.1016/S0004-3702(01)00106-0 Cited on page(s) 64, 65, 67

[106] Peter E. Hart, Nils J. Nilsson, and Bertram Raphael. A formal basis for the heuristic determination of minimum cost paths. *IEEE Transactions on Systems Science and Cybernetics*, 4(2):100?107, 1968. DOI: 10.1109/TSSC.1968.300136 Cited on page(s) 66

[107] Milos Hauskrecht and Branislav Kveton. Linear program approximations for factored continuous-state Markov decision processes. In *Advances in Neural Information Processing Systems*, 2003. Cited on page(s) 146

[108] Milos Hauskrecht, Nicolas Meuleau, Leslie Pack Kaelbling, Thomas Dean, and Craig Boutilier. Hierarchical solution of markov decision processes using macro-actions. In *Proceedings of the Fourteenth Conference on Uncertainty in Artificial Intelligence*, pages 220–229, 1998. Cited on page(s) 129

[109] Bernhard Hengst. Discovering hierarchy in reinforcement learning with HEXQ. In *Proceedings of the Nineteenth International Conference on Machine Learning*, pages 243–250, 2002. Cited on page(s) 138

[110] Natalia Hernandez-Gardiol and Leslie Pack Kaelbling. Envelope-based planning in relational MDPs. In *Advances in Neural Information Processing Systems*, 2003. Cited on page(s) 152

[111] Jesse Hoey, Robert St-Aubin, Alan Hu, and Craig Boutilier. SPUDD: Stochastic planning using decision diagrams. In *Proceedings of the Fifteenth Conference on Uncertainty in Artificial Intelligence*, pages 279–288, 1999. Cited on page(s) 88, 90

[112] Jörg Hoffmann and Bernhard Nebel. The FF planning system: Fast plan generation through heuristic search. *Journal of Artificial Intelligence Research*, 14:253–302, 2001.
DOI: 10.1613/jair.855 Cited on page(s) 99, 115

[113] Steffen Hölldobler, Eldar Karabaev, and Olga Skvortsova. Flucap: A heuristic search planner for first-order MDPs. *Journal of Artificial Intelligence Research*, 27:419–439, 2006.
DOI: 10.1613/jair.1965 Cited on page(s) 152

[114] Ronald A. Howard. *Dynamic Programming and Markov Processes*. MIT Press, 1960. Cited on page(s) 1, 36

[115] Ronald A. Howard. Semi-Markovian decision processes. In *Session International Statistical Institute*, pages 625–652, 1963. DOI: 10.1017/S026996480700037X Cited on page(s) 128, 148

[116] Ronald A. Howard. Comments on the origin and application of Markov decision processes. In Martin Puterman, editor, *Dynamic Programming and its Applications*, pages 201–205. Academic Press, New York, 1979. Cited on page(s) 3

[117] C. C. White III and H. K. El-Deib. Markov decision processes with imprecise transition probabilities. *Operations Research*, 42(4):739–749, 1994. DOI: 10.1287/opre.42.4.739 Cited on page(s) 158

[118] Anders Jonsson and Andrew G. Barto. Causal graph based decomposition of factored MDPs. *Journal of Machine Learning Research*, 7:2259–2301, 2006. Cited on page(s) 138

[119] Saket Joshi, Kristian Kersting, and Roni Khardon. Self-taught decision theoretic planning with first order decision diagrams. In *Proceedings of the Eighth International Conference on Automated Planning and Scheduling*, pages 89–96, 2010. Cited on page(s) 152

[120] Saket Joshi and Roni Khardon. Probabilistic relational planning with first order decision diagrams. *Journal of Artificial Intelligence Research*, 41:231–266, 2011. DOI: 10.1016/j.artint.2011.09.001 Cited on page(s) 94, 152

[121] Leslie P. Kaelbling, Michael L. Littman, and Andrew W. Moore. Reinforcement learning: A survey. *Journal of Artificial Intelligence Research*, 4:237–285, 1996. Cited on page(s) 4

[122] Leslie Pack Kaelbling. Hierarchical learning in stochastic domains: Preliminary results. In *Proceedings of the Tenth International Conference on Machine Learning*, pages 167–173, 1993. Cited on page(s) 135

[123] Leslie Pack Kaelbling, Michael L. Littman, and Anthony R. Cassandra. Planning and acting in partially observable stochastic domains. *Artificial Intelligence*, 101(1-2):99–134, 1998. DOI: 10.1016/S0004-3702(98)00023-X Cited on page(s) 145, 158

[124] Thomas Keller and Patrick Eyerich. Probabilistic planning based on UCT. In *Proceedings of the Tenth International Conference on Automated Planning and Scheduling*, 2012. Cited on page(s) 113

[125] Kristian Kersting, Martijn van Otterlo, and Luc De Raedt. Bellman goes relational. In *Proceedings of the Twenty-first International Conference on Machine Learning*, 2004. DOI: 10.1145/1015330.1015401 Cited on page(s) 152

[126] Emil Keyder and Hector Geffner. The HMDP planner for planning with probabilities. In *Sixth International Planning Competition at ICAPS'08*, 2008. Cited on page(s) 106, 107

[127] Craig A. Knoblock. Learning abstraction hierarchies for problem solving. In *Proceedings of the Seventh National Conference on Artificial Intelligence*, pages 923–928, 1990. Cited on page(s) 136

[128] László Kocsis and Csaba Szepesvári. Bandit based Monte-Carlo planning. In *Proceedings of the Seventeenth European Conference on Machine Learning*, pages 282–293, 2006. DOI: 10.1007/11871842_29 Cited on page(s) 111, 113

[129] Sven Koenig. Optimal probabilistic and decision-theoretic planning using Markovian decision theory. M.S. thesis UCB/CSD-92-685, University of California, Berkeley, 1991. Cited on page(s) 1

[130] Andrey Kolobov, Peng Dai, Mausam, and Daniel S. Weld. Reverse iterative deepening for finite-horizon MDPs with large branching factors. In *Proceedings of the Tenth International Conference on Automated Planning and Scheduling*, 2012. Cited on page(s) 71

[131] Andrey Kolobov, Mausam, and Daniel S. Weld. ReTrASE: Intergating paradigms for approximate probabilistic planning. In *Proceedings of the Twenty-first International Joint Conference on Artificial Intelligence*, 2009. Cited on page(s) 120

[132] Andrey Kolobov, Mausam, and Daniel S. Weld. Classical planning in MDP heuristics: with a little help from generalization. In *Proceedings of the Eighth International Conference on Automated Planning and Scheduling*, pages 97–104, 2010. Cited on page(s) 117

[133] Andrey Kolobov, Mausam, and Daniel S. Weld. SixthSense: Fast and reliable recognition of dead ends in MDPs. In *Proceedings of the Twenty-fourth AAAI Conference on Artificial Intelligence*, 2010. Cited on page(s) 110, 156, 157

[134] Andrey Kolobov, Mausam, and Daniel S. Weld. Discovering hidden structure in factored MDPs. *Artificial Intelligence*, 189:19–47, 2012. DOI: 10.1016/j.artint.2012.05.002 Cited on page(s) 117, 120, 156, 157

[135] Andrey Kolobov, Mausam, and Daniel S. Weld. LRTDP vs. UCT for online probabilistic planning. In *Proceedings of the Twenty-sixth AAAI Conference on Artificial Intelligence*, 2012. Cited on page(s) 71

[136] Andrey Kolobov, Mausam, and Daniel S. Weld. Stochastic shortest path MDPs with dead ends. In *ICAPS Heuristics and Search for Domain Independent Planning (HSDIP) Workshop*, 2012. Cited on page(s) 57, 58, 81, 153, 156, 157

[137] Andrey Kolobov, Mausam, Daniel S. Weld, and Hector Geffner. Heuristic search for generalized stochastic shortest path MDPs. In *Proceedings of the Ninth International Conference on Automated Planning and Scheduling*, 2011. Cited on page(s) 108, 110, 153, 156, 157

[138] George Konidaris, Scott Kuindersma, Roderic A. Grupen, and Andrew G. Barto. Autonomous skill acquisition on a mobile manipulator. In *Proceedings of the Twenty-fifth AAAI Conference on Artificial Intelligence*, 2011. Cited on page(s) 130

[139] Richard Korf. Real-time heuristic search. *Artificial Intelligence*, 2-3:189?211, 1990. DOI: 10.1016/0004-3702(90)90054-4 Cited on page(s) 78

[140] Ugur Kuter and Jiaqiao Hu. Computing and using lower and upper bounds for action elimination in MDP planning. In *Proceedings of the 7th International conference on Abstraction, reformulation, and approximation*, 2007. DOI: 10.1007/978-3-540-73580-9_20 Cited on page(s) 76

[141] Ugur Kuter and Dana S. Nau. Using domain-configurable search control for probabilistic planning. In *Proceedings of the Twentieth National Conference on Artificial Intelligence*, pages 1169–1174, 2005. Cited on page(s) 136

[142] Branislav Kveton and Milos Hauskrecht. Heuristic refinements of approximate linear programming for factored continuous-state Markov decision processes. In *Proceedings of the Second International Conference on Automated Planning and Scheduling*, pages 306–314, 2004. Cited on page(s) 146

[143] Branislav Kveton and Milos Hauskrecht. An MCMC approach to solving hybrid factored MDPs. In *Proceedings of the Nineteenth International Joint Conference on Artificial Intelligence*, pages 1346–1351, 2005. Cited on page(s) 146

[144] Branislav Kveton and Milos Hauskrecht. Solving factored MDPs with exponential-family transition models. In *Proceedings of the Fourth International Conference on Automated Planning and Scheduling*, pages 114–120, 2006. Cited on page(s) 146

[145] Branislav Kveton, Milos Hauskrecht, and Carlos Guestrin. Solving factored MDPs with hybrid state and action variables. *Journal of Artificial Intelligence Research*, 27:153–201, 2006. DOI: 10.1613/jair.2085 Cited on page(s) 146, 147

[146] Michail G. Lagoudakis and Ronald Parr. Least-squares policy iteration. *Journal of Machine Learning Research*, 4:1107–1149, 2003. Cited on page(s) 146

[147] Anthony LaMarca and Richard E. Ladner. The influence of caches on the performance sorting. *Journal of Algorithms*, 31:66–104, 1999. DOI: 10.1006/jagm.1998.0985 Cited on page(s) 53

[148] Frank L. Lewis, Daguna Vrabie, and Vassilis L. Syrmos. *Optimal Control.* Wiley, 2012. DOI: 10.1002/9781118122631 Cited on page(s) 3, 147

[149] Li Li and Nilufer Onder. Generating plans in concurrent, probabilistic, over-subscribed domains. In *Proceedings of the Twenty-third AAAI Conference on Artificial Intelligence*, pages 957–962, 2008. Cited on page(s) 150

[150] Lihong Li and Michael L. Littman. Lazy approximation for solving continuous finite-horizon MDPs. In *Proceedings of the Twentieth National Conference on Artificial Intelligence*, pages 1175–1180, 2005. Cited on page(s) 145

[151] Lihong Li and Michael L. Littman. Prioritized sweeping converges to the optimal value function. Technical Report DCS-TR-631, Rutgers University, 2008. Cited on page(s) 45, 46, 47

[152] Iain Little, Douglas Aberdeen, and Sylvie Thiébaux. Prottle: A probabilistic temporal planner. In *Proceedings of the Twentieth National Conference on Artificial Intelligence*, pages 1181–1186, 2005. Cited on page(s) 150

[153] Iain Little and Sylvie Thiébaux. Concurrent probabilistic planning in the Graphplan framework. In *Proceedings of the Fourth International Conference on Automated Planning and Scheduling*, pages 263–273, 2006. Cited on page(s) 149

[154] Iain Little and Sylvie Thiebaux. Probabilistic planning vs. replanning. In *ICAPS Workshop on IPC: Past, Present and Future*, 2007. Cited on page(s) 109

[155] Michael L. Littman. Markov games as a framework for multi-agent reinforcement learning. In *Proceedings of the Eleventh International Conference on Machine Learning*, pages 157–163, 1994. Cited on page(s) 159

[156] Michael L. Littman. Probabilistic propositional planning: representations and complexity. In *Proceedings of the Fourteenth National Conference on Artificial Intelligence*, 1997. Cited on page(s) 28, 29

[157] Michael L. Littman, Thomas L. Dean, and Leslie Pack Kaelbling. On the complexity of solving Markov decision problems. In *Proceedings of the Eleventh Conference on Uncertainty in Artificial Intelligence*, pages 394–402, 1995. Cited on page(s) 55

[158] Yaxin Liu and Sven Koenig. Risk-sensitive planning with one-switch utility functions: Value iteration. In *Proceedings of the Twentieth National Conference on Artificial Intelligence*, pages 993–999, 2005. Cited on page(s) 146

[159] Yaxin Liu and Sven Koenig. Functional value iteration for decision-theoretic planning with general utility functions. In *Proceedings of the Twenty-first National Conference on Artificial Intelligence*, 2006. Cited on page(s) 145

[160] Yaxin Liu and Sven Koenig. An exact algorithm for solving MDPs under risk-sensitive planning objectives with one-switch utility functions. In *Proceedings of the Seventh International Conference on Autonomous Agents and Multiagent Systems*, pages 453–460, 2008. DOI: 10.1145/1402383.1402449 Cited on page(s) 146

[161] Pattie Maes and Rodney A. Brooks. Learning to coordinate behaviors. In *Proceedings of the Seventh National Conference on Artificial Intelligence*, pages 796–802, 1990. Cited on page(s) 135

[162] Janusz Marecki, Sven Koenig, and Milind Tambe. A fast analytical algorithm for solving Markov decision processes with real-valued resources. In *Proceedings of the Twentieth International Joint Conference on Artificial Intelligence*, pages 2536–2541, 2007. Cited on page(s) 146

[163] Janusz Marecki and Milind Tambe. Towards faster planning with continuous resources in stochastic domains. In *Proceedings of the Twenty-third AAAI Conference on Artificial Intelligence*, pages 1049–1055, 2008. Cited on page(s) 147

[164] Bhaskara Marthi, Leslie Pack Kaelbling, and Tomas Lozano-Perez. Learning hierarchical structure in policies. In *NIPS Hierarchical Organization of Behavior Workshop*, 2007. Cited on page(s) 137, 138

[165] Mausam. Stochastic planning with concurrent, durative actions. Ph.D. thesis, University of Washington, Seattle, 2007. Cited on page(s) 141

[166] Mausam, Emmanuelle Benazara, Ronen Brafman, Nicolas Meuleau, and Eric A. Hansen. Planning with continuous resources in stochastic domains. In *Proceedings of the Nineteenth International Joint Conference on Artificial Intelligence*, pages 1244–1251, 2005. Cited on page(s) 1, 144, 146, 147

[167] Mausam, Piergiorgio Bertoli, and Daniel S. Weld. A hybridized planner for stochastic domains. In *Proceedings of the Twentieth International Joint Conference on Artificial Intelligence*, pages 1972–1978, 2007. Cited on page(s) 140, 141

[168] Mausam and Daniel S. Weld. Solving relational MDPs with first-order machine learning. In *Proceedings of ICAPS Workshop on Planning under Uncertainty and Incomplete Information*, 2003. Cited on page(s) 151, 152

[169] Mausam and Daniel S. Weld. Solving concurrent Markov decision processes. In *Proceedings of the Nineteenth National Conference on Artificial Intelligence*, pages 716–722, 2004. Cited on page(s) 149

[170] Mausam and Daniel S. Weld. Concurrent probabilistic temporal planning. In *Proceedings of the Third International Conference on Automated Planning and Scheduling*, pages 120–129, 2005. Cited on page(s) 149

[171] Mausam and Daniel S. Weld. Challenges for temporal planning with uncertain durations. In *Proceedings of the Fourth International Conference on Automated Planning and Scheduling*, pages 414–417, 2006. Cited on page(s) 150

[172] Mausam and Daniel S. Weld. Probabilistic temporal planning with uncertain durations. In *Proceedings of the Twenty-first National Conference on Artificial Intelligence*, 2006. Cited on page(s) 150

[173] Mausam and Daniel S. Weld. Planning with durative actions in stochastic domains. *Journal of Artificial Intelligence Research*, 31:33–82, 2008. DOI: 10.1613/jair.2269 Cited on page(s) 148, 149, 150

[174] Amy McGovern and Andrew G. Barto. Automatic discovery of subgoals in reinforcement learning using diverse density. In *Proceedings of the Eighteenth International Conference on Machine Learning*, pages 361–368, 2001. Cited on page(s) 137

[175] H. Brendan McMahan and Geoffrey J. Gordon. Fast exact planning in Markov decision processes. In *Proceedings of the Third International Conference on Automated Planning and Scheduling*, pages 151–160, 2005. Cited on page(s) 47

[176] H. Brendan Mcmahan, Maxim Likhachev, and Geoffrey J. Gordon. Bounded real-time dynamic programming: RTDP with monotone upper bounds and performance guarantees. In *Proceedings of the Twenty-second International Conference on Machine Learning*, pages 569–576, 2005. DOI: 10.1145/1102351.1102423 Cited on page(s) 72, 74, 76

[177] Neville Mehta, Soumya Ray, Prasad Tadepalli, and Thomas G. Dietterich. Automatic discovery and transfer of MAXQ hierarchies. In *Proceedings of the Twenty-fifth International Conference on Machine Learning*, pages 648–655, 2008. DOI: 10.1145/1390156.1390238 Cited on page(s) 138

[178] Ishai Menache, Shie Mannor, and Nahum Shimkin. Q-cut - dynamic discovery of sub-goals in reinforcement learning. In *Proceedings of the Thirteenth European Conference on Machine Learning*, pages 295–306, 2002. Cited on page(s) 137

[179] Nicolas Meuleau, Emmanuel Benazera, Ronen I. Brafman, Eric A. Hansen, and Mausam. A heuristic search approach to planning with continuous resources in stochastic domains. *Journal of Artificial Intelligence Research*, 34:27–59, 2009. DOI: 10.1613/jair.2529 Cited on page(s) 146, 147

[180] Nicolas Meuleau and Ronen I. Brafman. Hierarchical heuristic forward search in stochastic domains. In *Proceedings of the Twentieth International Joint Conference on Artificial Intelligence*, pages 2542–2549, 2007. Cited on page(s) 136

[181] Nicolas Meuleau, Milos Hauskrecht, Kee-Eung Kim, Leonid Peshkin, Leslie Kaelbling, Thomas Dean, and Craig Boutilier. Solving very large weakly coupled Markov Decision Processes. In *Proceedings of the Fifteenth National Conference on Artificial Intelligence*, pages 165–172, 1998. Cited on page(s) 149

[182] Nicolas Meuleau and David Smith. Optimal limited contingency planning. In *Proceedings of the Nineteenth Conference on Uncertainty in Artificial Intelligence*, pages 417–426, 2003. Cited on page(s) 148

[183] Andrew W. Moore and Christopher G. Atkeson. Prioritized sweeping: Reinforcement learning with less data and less time. *Machine Learning*, 13:103–130, 1993. DOI: 10.1007/BF00993104 Cited on page(s) 46

[184] Andrew W. Moore and Christopher G. Atkeson. The parti-game algorithm for variable resolution reinforcement learning in multidimensional state space. *Machine Learning*, 21:199–233, 1995. DOI: 10.1023/A:1022656217772 Cited on page(s) 51

[185] Andrew W. Moore, Leemon C. Baird III, and Leslie Pack Kaelbling. Multi-value-functions: Efficient automatic action hierarchies for multiple goal MDPs. In *Proceedings of the Sixteenth International Joint Conference on Artificial Intelligence*, pages 1316–1323, 1999. Cited on page(s) 135

[186] Remi Munos and Andrew W. Moore. Variable resolution discretization in optimal control. *Machine Learning*, 49:291–323, 2002. DOI: 10.1023/A:1017992615625 Cited on page(s) 51, 145

[187] I. Murthy and S. Sarkar. Stochastic shortest path problems with piecewise-linear concave utility functions. *Management Science*, 44(11):125–136, 1998. DOI: 10.1287/mnsc.44.11.S125 Cited on page(s) 145

[188] Dana S. Nau, Tsz-Chiu Au, Okhtay Ilghami, Ugur Kuter, J. William Murdock, Dan Wu, and Fusun Yaman. Shop2: An HTN planning system. *Journal of Artificial Intelligence Research*, 20:379–404, 2003. DOI: 10.1613/jair.1141 Cited on page(s) 136

[189] Nils J. Nilsson. *Principles of Artificial Intelligence*. Tioga Publishing, 1980. Cited on page(s) 68

[190] Frans Oliehoek. Decentralized POMDPs. In *Reinforcement Learning: State of the Art*. Springer, Berlin, 2011. Cited on page(s) 159

[191] Christos H. Papadimitriou and John N. Tsitsiklis. The complexity of Markov decision processes. *Mathematics of Operations Research*, 12(3):441–450, 1987. DOI: 10.1287/moor.12.3.441 Cited on page(s) 28

[192] Ronald Parr and Stuart J. Russell. Reinforcement learning with hierarchies of machines. In *Advances in Neural Information Processing Systems*, 1997. Cited on page(s) 133

[193] Relu Patrascu, Pascal Poupart, Dale Schuurmans, Craig Boutilier, and Carlos Guestrin. Greedy linear value-approximation for factored Markov decision processes. In *Proceedings of the Eighteenth National Conference on Artificial Intelligence*, pages 285–291, 2002. Cited on page(s) 145

[194] Joelle Pineau, Geoffrey J. Gordon, and Sebastian Thrun. Policy-contingent abstraction for robust robot control. In *Proceedings of the Nineteenth Conference on Uncertainty in Artificial Intelligence*, pages 477–484, 2003. Cited on page(s) 137

[195] Jeffrey L. Popyack. Blackjack-playing agents in an advanced ai course. In *Proceedings of the Fourteenth Annual SIGCSE Conference on Innovation and Technology in Computer Science Education*, pages 208–212, 2009. DOI: 10.1145/1562877.1562944 Cited on page(s) 1

[196] Pascal Poupart, Craig Boutilier, Relu Patrascu, and Dale Schuurmans. Piecewise linear value function approximation for factored MDPs. In *Proceedings of the Eighteenth National Conference on Artificial Intelligence*, pages 292–299, 2002. Cited on page(s) 145

[197] Martin L. Puterman. *Markov Decision Processes*. John Wiley & Sons, 1994. DOI: 10.1002/9780470316887 Cited on page(s) 4, 15, 17, 36, 54, 55, 128, 148, 153

[198] Martin L. Puterman and M. C. Shin. Modified policy iteration algorithms for discounted Markov decision problems. *Management Science*, 24, 1978. DOI: 10.1287/mnsc.24.11.1127 Cited on page(s) 38

[199] Emmanuel Rachelson, Frederick Garcia, and Patrick Fabiani. Extending the bellman equation for MDPs to continuous actions and continuous time in the discounted case. In *Proceedings of Tenth International Symposium on Artificial Intelligence and Mathematics*, 2008. Cited on page(s) 147

[200] Anna N. Rafferty, Emma Brunskill, Thomas L. Griffiths, and Patrick Shafto. Faster teaching by POMDP planning. In *Proceedings of Artificial Intelligence in Education*, pages 280–287, 2011. DOI: 10.1007/978-3-642-21869-9_37 Cited on page(s) 159

[201] Aswin Raghavan, Saket Joshi, Alan Fern, Prasad Tadepalli, and Roni Khardon. Planning in factored action spaces with symbolic dynamic programming. In *Proceedings of the Twenty-sixth AAAI Conference on Artificial Intelligence*, 2012. Cited on page(s) 149

[202] Rajesh P. N. Rao. Decision making under uncertainty: A neural model based on partially observable Markov decision processes. *Frontiers in Computational Neuroscience*, 4, 2010. DOI: 10.3389/fncom.2010.00146 Cited on page(s) 3

[203] Silvia Richter and Matthias Westphal. The LAMA planner: Guiding cost-based anytime planning with landmarks. *Journal of Artificial Intelligence Research*, 39:127–177, 2010. DOI: 10.1613/jair.2972 Cited on page(s) 99

[204] Scott Sanner. ICAPS 2011 international probabilistic planning competition. `http://users.cecs.anu.edu.au/~ssanner/IPPC_2011/`, 2011. Cited on page(s) 15, 16, 26, 71, 110, 113

[205] Scott Sanner and Craig Boutilier. Approximate linear programming for first-order MDPs. In *Proceedings of the Twenty-first Conference on Uncertainty in Artificial Intelligence*, pages 509–517, 2005. Cited on page(s) 152

[206] Scott Sanner and Craig Boutilier. Practical linear value-approximation techniques for first-order MDPs. In *Proceedings of the Twenty-second Conference on Uncertainty in Artificial Intelligence*, 2006. Cited on page(s) 152

[207] Scott Sanner and Craig Boutilier. Approximate solution techniques for factored first-order MDPs. In *Proceedings of the Fifth International Conference on Automated Planning and Scheduling*, pages 288–295, 2007. Cited on page(s) 151

[208] Scott Sanner and Craig Boutilier. Practical solution techniques for first-order MDPs. *Artificial Intelligence*, 173(5-6):748–788, 2009. DOI: 10.1016/j.artint.2008.11.003 Cited on page(s) 152

[209] Scott Sanner, Karina Valdivia Delgado, and Leliane Nunes de Barros. Symbolic dynamic programming for discrete and continuous state MDPs. In *Proceedings of the Twenty-seventh Conference on Uncertainty in Artificial Intelligence*, pages 643–652, 2011. Cited on page(s) 94, 143, 146

[210] Scott Sanner, Robby Goetschalckx, Kurt Driessens, and Guy Shani. Bayesian real-time dynamic programming. In *Proceedings of the Twenty-first International Joint Conference on Artificial Intelligence*, 2009. Cited on page(s) 74

[211] Scott Sanner and David A. McAllestor. Affine algebraic decision diagrams (AADDs) and their application to structured probabilistic inference. In *Proceedings of the Nineteenth International Joint Conference on Artificial Intelligence*, pages 1384–1390, 2005. Cited on page(s) 93, 94

[212] Scott Sanner, William T. B. Uther, and Karina Valdivia Delgado. Approximate dynamic programming with affine ADDs. In *Proceedings of the Ninth International Conference on Autonomous Agents and Multiagent Systems*, pages 1349–1356, 2010. DOI: 10.1145/1838206.1838383 Cited on page(s) 95

182 BIBLIOGRAPHY

[213] J. K. Satia and R. E. Lave Jr. Markov decision processes with uncertain transition probabilities. *Operations Research*, 21:728–740, 1970. DOI: 10.1016/j.fss.2004.10.023 Cited on page(s) 158

[214] Guy Shani, Pascal Poupart, Ronen I. Brafman, and Solomon Eyal Shimony. Efficient ADD operations for point-based algorithms. In *Proceedings of the Sixth International Conference on Automated Planning and Scheduling*, pages 330–337, 2008. Cited on page(s) 159

[215] Özgür Simsek and Andrew G. Barto. Using relative novelty to identify useful temporal abstractions in reinforcement learning. In *Proceedings of the Twenty-first International Conference on Machine Learning*, 2004. DOI: 10.1145/1015330.1015353 Cited on page(s) 137

[216] Özgür Simsek and Andrew G. Barto. Skill characterization based on betweenness. In *Advances in Neural Information Processing Systems*, pages 1497–1504, 2008. Cited on page(s) 137

[217] Satinder Singh and David Cohn. How to dynamically merge Markov decision processes. In *Advances in Neural Information Processing Systems*, 1998. Cited on page(s) 149

[218] Satinder P. Singh. Reinforcement learning with a hierarchy of abstract models. In *Proceedings of the Ninth National Conference on Artificial Intelligence*, pages 202–207, 1992. Cited on page(s) 135

[219] Trey Smith and Reid G. Simmons. Focused real-time dynamic programming for MDPs: Squeezing more out of a heuristic. In *Proceedings of the Twenty-first National Conference on Artificial Intelligence*, 2006. Cited on page(s) 74

[220] Fabio Somenzi. CUDD: CU decision diagram package, url: http://vlsi.colorado.edu/~fabio/CUDD/, 1998. Cited on page(s) 87

[221] Robert St-Aubin, Jesse Hoey, and Craig Boutilier. APRICODD: Approximate policy construction using decision diagrams. In *Advances in Neural Information Processing Systems*, pages 1089–1095, 2000. Cited on page(s) 95

[222] Robert F. Stengel. *Optimal Control and Estimation*. Dover Publications, 1994. Cited on page(s) 3, 147

[223] Peter Stone, Richard S. Sutton, and Gregory Kuhlmann. Reinforcement learning for RoboCup-soccer keepaway. *Adaptive Control*, 2005. DOI: 10.1177/105971230501300301 Cited on page(s) 1

[224] Richard S. Sutton and Andrew G. Barto. *Reinforcement Learning: An Introduction*. MIT Press, 1998. Cited on page(s) 4, 158

[225] Richard S. Sutton, Doina Precup, and Satinder P. Singh. Between MDPs and semi-MDPs: A framework for temporal abstraction in reinforcement learning. *Artificial Intelligence*, 112(1-2):181–211, 1999. DOI: 10.1016/S0004-3702(99)00052-1 Cited on page(s) 127

[226] Csaba Szepesvári. *Algorithms for Reinforcement Learning*. Synthesis Lectures on Artificial Intelligence and Machine Learning. Morgan & Claypool Publishers, 2010. DOI: 10.2200/S00268ED1V01Y201005AIM009 Cited on page(s) 4, 158

[227] Yuqing Tang, Felipe Meneguzzi, Katia Sycara, and Simon Parsons. Planning over MDPs through probabilistic HTNs. In *AAAI 2011 Workshop on Generalized Planning*, 2011. Cited on page(s) 136

[228] Florent Teichteil-Königsbuch, Ugur Kuter, and Guillaume Infantes. Incremental plan aggregation for generating policies in MDPs. In *Proceedings of the Ninth International Conference on Autonomous Agents and Multiagent Systems*, pages 1231–1238, 2010. DOI: 10.1145/1838206.1838366 Cited on page(s) 103, 110

[229] Florent Teichteil-Königsbuch, Vincent Vidal, and Guillaume Infantes. Extending classical planning heuristics to probabilistic planning with dead-ends. In *Proceedings of the Twenty-fifth AAAI Conference on Artificial Intelligence*, 2011. Cited on page(s) 78

[230] Florent Teichteil-Königsbuch. Fast incremental policy compilation from plans in hybrid probabilistic domains. In *Proceedings of the Tenth International Conference on Automated Planning and Scheduling*, 2012. Cited on page(s) 146

[231] Florent Teichteil-Königsbuch. Path-constrained Markov decision processes: bridging the gap between probabilistic model-checking and decision-theoretic planning. In *Proceedings of the Twentieth European Conference on Artificial Intelligence*, 2012. Cited on page(s) 156

[232] Florent Teichteil-Königsbuch. Stochastic safest and shortest path problems. In *Proceedings of the Twenty-sixth AAAI Conference on Artificial Intelligence*, 2012. Cited on page(s) 156

[233] Sebastian Thrun, Michael Beetz, Maren Bennewitz, Wolfram Burgard, Armin B. Cremers, Frank Dellaert, Dieter Fox, Dirk Hähnel, Charles R. Rosenberg, Nicholas Roy, Jamieson Schulte, and Dirk Schulz. Probabilistic algorithms and the interactive museum tour-guide robot Minerva. *International Journal of Robotic Research*, 19(11):972–999, 2000. DOI: 10.1177/02783640022067922 Cited on page(s) 159

[234] Margaret S. Trench, Shane P. Pederson, Edward T. Lau, Lizhi Ma, Hui Wang, and Suresh K. Nair. Managing credit lines and prices for bank one credit cards. *Interfaces*, 33:4–21, 2003. DOI: 10.1287/inte.33.5.4.19245 Cited on page(s) 3

[235] Felipe W. Trevizan, Fabio Gagliardi Cozman, and Leliane Nunes de Barros. Planning under risk and knightian uncertainty. In *Proceedings of the Twentieth International Joint Conference on Artificial Intelligence*, pages 2023–2028, 2007. Cited on page(s) 153, 158

[236] P. Tseng. Solving h horizon stationary Markov decision problems in time proportional to $\log(h)$. *Operations Research Letters*, 9(5):287–297, 1990. DOI: 10.1016/0167-6377(90)90022-W Cited on page(s) 55

[237] J. A. E. E. van Nunen. A set of successive approximation methods for discounted Markovian decision problems. *Mathematical Methods of Operations Research*, 20(5):203–208, 1976. DOI: 10.1007/BF01920264 Cited on page(s) 38

[238] K. Wakuta. Vector-valued Markov decisionprocesses and the systems of linear inequalities. *Stochastic Processes and their Applications*, 56:159–169, 1995. DOI: 10.1016/0304-4149(94)00064-Z Cited on page(s) 157

[239] Chenggang Wang and Roni Khardon. Policy iteration for relational MDPs. In *UAI*, pages 408–415, 2007. Cited on page(s) 152

[240] Douglas J. White. Real applications of Markov decision processes. *Interfaces*, 15(6):73–83, 1985. DOI: 10.1287/inte.15.6.73 Cited on page(s) 3

[241] Jason D. Williams and Steve Young. Partially observable Markov decision processes for spoken dialog systems. *Computer Speech & Language*, 21(2):393–422, 2007. DOI: 10.1016/j.csl.2006.06.008 Cited on page(s) 159

[242] Ron J. Williams and Leemon Baird. Tight performance bounds on greedy policies based on imperfect value functions. In *Proceedings of the Tenth Yale Workshop on Adaptive and Learning Systems*, 1994. Cited on page(s) 54

[243] David Wingate and Kevin D. Seppi. Efficient value iteration using partitioned models. In *Proceedings of the International Conference on Machine Learning and Applications*, pages 53–59, 2003. Cited on page(s) 50

[244] David Wingate and Kevin D. Seppi. Cache performance of priority metrics for MDP solvers. In *Proceedings of AAAI Workshop on Learning and Planning in Markov Processes*, pages 103–106, 2004. Cited on page(s) 53

[245] David Wingate and Kevin D. Seppi. P3VI: a partitioned, prioritized, parallel value iterator. In *Proceedings of the Twenty-first International Conference on Machine Learning*, 2004. Cited on page(s) 53

[246] David Wingate and Kevin D. Seppi. Prioritization methods for accelerating MDP solvers. *Journal of Machine Learning Research*, 6:851–881, 2005. Cited on page(s) 45, 48, 50, 55

[247] Jia-Hong Wu and Robert Givan. Automatic induction of bellman-error features for probabilistic planning. *Journal of Artificial Intelligence Research*, 38:687–755, 2010. Cited on page(s) 152

[248] Sung Wook Yoon, Alan Fern, and Robert Givan. Inductive policy selection for first-order MDPs. In *Proceedings of the Eighteenth Conference on Uncertainty in Artificial Intelligence*, pages 568–576, 2002. Cited on page(s) 152

[249] Sung Wook Yoon, Wheeler Ruml, J. Benton, and Minh Binh Do. Improving determinization in hindsight for on-line probabilistic planning. In *Proceedings of the Eighth International Conference on Automated Planning and Scheduling*, pages 209–217, 2010. Cited on page(s) 103

[250] Sungwook Yoon, Alan Fern, and Robert Givan. FF-Replan: A baseline for probabilistic planning. In *Proceedings of the Fifth International Conference on Automated Planning and Scheduling*, 2007. Cited on page(s) 77, 100, 101

[251] Sungwook Yoon, Alan Fern, Subbarao Kambhampati, and Robert Givan. Probabilistic planning via determinization in hindsight. In *Proceedings of the Twenty-third AAAI Conference on Artificial Intelligence*, pages 1010–1016, 2008. Cited on page(s) 102

[252] Håkan L. S. Younes and Michael Littman. PPDDL1.0: The language for the probabilistic part of IPC-4. In *Fourth International Planning Competition at ICAPS'04*, 2004. Cited on page(s) 26

[253] Hakan L. S. Younes and Michael L. Littman. PPDDL1.0: An extension to PDDL for expressiong planning domains with probabilistic effects. Technical Report CMU-CS-04-167, Carnegie Mellon University, 2004. Cited on page(s) 52, 151

[254] Håkan L. S. Younes, Michael L. Littman, David Weissman, and John Asmuth. The first probabilistic track of the international planning competition. *Journal of Artificial Intelligence Research*, 24:851–887, 2005. Cited on page(s) 91

[255] Håkan L. S. Younes and Reid G. Simmons. Policy generation for continuous-time stochastic domains with concurrency. In *ICAPS*, pages 325–334, 2004. Cited on page(s) 150

[256] Håkan L. S. Younes and Reid G. Simmons. Solving generalized semi-markov decision processes using continuous phase-type distributions. In *Proceedings of the Nineteenth National Conference on Artificial Intelligence*, pages 742–748, 2004. Cited on page(s) 150

[257] Zahra Zamani, Scott Sanner, and Cheng Fang. Symbolic dynamic programming for continuous state and action MDPs. In *Proceedings of the Twenty-sixth AAAI Conference on Artificial Intelligence*, 2012. Cited on page(s) 147

Authors' Biographies

MAUSAM

Mausam is a Research Assistant Professor at the Turing Center in the Department of Computer Science at the University of Washington, Seattle. His research interests span various sub-fields of artificial intelligence, including sequential decision making under uncertainty, large scale natural language processing, Web information systems, heuristic search, machine learning, and AI applications to crowd-sourcing. Mausam obtained a Ph.D. from University of Washington in 2007 and a Bachelor of Technology from IIT Delhi in 2001. His PhD thesis was awarded honorable mention for the 2008 ICAPS Best Dissertation Award. Mausam has written scores of papers in top AI conferences and journals. He has served on the senior program committees of AI conferences such as AAAI and IJCAI, program committees of several other conferences, and on NSF panels.

ANDREY KOLOBOV

Andrey Kolobov is a Ph.D. student at the Department of Computer Science at the University of Washington, Seattle, advised by Professors Mausam and Daniel S. Weld. His research concentrates primarily on designing scalable domain-independent algorithms for planning under uncertainty, but has touched upon other areas as well, including information retrieval, robotics, first-order probabilistic languages, and transfer learning. Andrey has authored multiple works in top AI venues, including a publication in the exceptional paper track at AAAI-2010, the best paper at the Doctoral Consortium at ICAPS-2009, and a runner-up entry at the International Probabilistic Planning Competition-2011. He is also a recipient of the Outstanding Program Committee Member Award at AAAI-2012.

Index

Printed in the United States
by Baker & Taylor Publisher Services